Johann Georg Zimmermann

A Treatise on the Dysentery

With a Description of the Epidemic Dysentery that Happened in Switzerland

Johann Georg Zimmermann

A Treatise on the Dysentery
With a Description of the Epidemic Dysentery that Happened in Switzerland

ISBN/EAN: 9783337156183

Printed in Europe, USA, Canada, Australia, Japan

Cover: Foto ©berggeist007 / pixelio.de

More available books at **www.hansebooks.com**

A

TREATISE

ON THE

DYSENTERY:

With a DESCRIPTION of the

EPIDEMIC DYSENTE

That happened in SWITZERLAND
Year 1765.

Tranflated from the Original GERMA

JOHN GEORGE ZIMMERMAN,

Phyfician in Ordinary to his BRITANNIC M
Hanover,

By C. R. HOPSON,

LONDON:

Printed for JOHN and FRANCIS RI
at the Bible and Crown (Nº 6
St. Paul's Church-yard.

M DCC

T O

WILLIAM CULLEN, M. D.

PROFESSOR of MEDICINE in the
Univerſity of EDINBURGH.

SIR,

THE diſtinguiſhing marks of
civility, which I received at
your hands during the ſtay I made
ſome time ſince at Edinburgh, have
induced me to offer the following
tranſlation, as a ſmall token of the
reſpect I owe you on many accounts.
Indeed I could not ſo properly de-
dicate it to any one in Britain as to
yourſelf: for (not to mention the
juſt claim you have to my acknow-
ledgements for the opportunities I
have enjoyed of improving in my
profeſſion under your auſpices) you
are likewiſe entitled to the thanks

A 2   of

DEDICATION.

of the author, as having already in the moſt effectual and diſtinguiſhed manner patronized his work, while it was in its native dreſs, by recommending it publicly to your diſciples: and if it ſhould now make its way in the world in its preſent diſguiſe, it will perhaps be more owing to the force of your recommendation, than to its own merits; which, however great, might have been otherwiſe totally obſcured by the defects of the tranſlator. Theſe defects, however, I humbly preſume, lie chiefly in the ſtile; and as they could not be well avoided (the tranſlation having been made ſeveral years ago in the midſt of Germany, when the tranſlator had poſſibly, by a long abode in foreign countries, ſubſtituted the idioms of their languages for thoſe of his mother tongue) it is to be hoped, they will find excuſe with you and the public.

Indeed

DEDICATION.

Indeed what I have principally aimed at in this tranflation is not elegance, but exactnefs; the former being by no means neceffary in a work of this nature: though the author has difplayed in this treatife a purity and elegance of ftile, which have contributed to place him in the firft clafs of German writers; and indeed has frequently indulged himfelf (agreeably to the genius of his countrymen) in fuch high flights of metaphorical expreffion and poetical language, as I have not in general ventured to imitate. But to do real juftice to his work, would require much more leifure, than my prefent occupations will allow: I have only been able to make a curfory review of it, and for the reft muft rely on your patronage for its favourable reception with the public; with which it will have the additional merit of being the only treatife, befides

DEDICATION.

fides your excellent Nofology, that
has ever eftablifhed the true crite-
rion of the diforder, the fubject of
the following pages; which are here
fubmitted to your indulgent infpec-
tion, with the greateft deference
and refpect, by,

S I R,

Your obliged

Humble fervant,

C. R. Hopson.

PREFACE.

THE following treatife was written originally in the German language by Dr. Zimmerman, a Swifs phyfician, a friend and difciple of the great Haller, and not more admired by his countrymen for the beauty of his ftile and force of expreffion, than for his great infight into the theory and practice of phyfic. He is the author of an univerfally admired work on *Experience in Medicine*, of which he has already publifhed two volumes, and has promifed a continuation, that is impatiently expected by the public. The Germans are fo fenfible of his merit, that fome time before the publication of the abovementioned production he was invited to fill the practical chair in his Majefty's univerfity of Gottingen, which honour he however declined. The work, of which the following pages are a tranflation, was chiefly intended by its public-fpirited au-

thor

thor to diffipate the prejudices of his countrymen (especially those of the lower clafs)
with regard to the treatment to be ufed in
the dyfentery. That part of it the tranflator
has omitted; and has occafionally abridged
fome other parts of the work, which did
not immediately refpect the principal fubject. As to the ftile, the tranflator is very
fenfible of his defects in that point, owing
principally to his abfence from his native
country; but hopes for indulgence from
the public, to which he was ambitious of
being ferviceable in any refpect.

T H E

CONTENTS.

CONTENTS.

C H A P. VI.

C H A P. VII.

A

A

TREATISE

ON THE

DYSENTERY.

CHAP. I.

Defcription of the DISORDER.

THE dyfentery has raged this year in the canton of Bern, in the marquifate of Torgau, in divers other parts of Switzerland, and in Swabia. It has been alfo very formidable in thofe diftricts of Upper-Auftria, which lie neareft to us.

It made its firft appearance in the month of June; in Auguft and September rofe to its higheft pitch; in the beginning of October loft ground in all parts; and, in the middle of this month, generally fpeaking, made its exit. Though, in the middle of November, fome here and there were feized with this diforder; and even during the fevereft cold in December,

B   and

and January 1766, I faw people who were attacked by a gentle dyfentery: in like manner,
about this time, under the fame conftitution
of the air, the putrid fever, as it is called among
us, and particularly the putrid pleurify, began
their ravages, principally in Laufanne, and extended them wide around as far as our canton,
and the neighbouring provinces of Upper-Auftria and Swabia.

Many were taken with this malady, without
the leaft preceding fymptom, and that chiefly
in defperate cafes; in others, it gave tokens
of its approach before hand, and came on by
degrees.

All thofe who were violently difordered, were
feized at firft with an univerfal chill, which had
different degrees of duration; fometimes it was
long and very violent, many had only a fmall
paroxyfm, with many it returned in the courfe
of the fever, and went off in a hot fit. All of
them felt likewife an extreme laffitude over the
whole body, at the firft coming on of the diforder, and that generally in the back and loins.
The cholic came on immediately at the beginning, with great violence; but the ▪cuation,
with fome people, did not follow fo quick;
many were at firft even bound, thefe had violent

lent tormina, and were in a much worfe condition, than thofe who were obliged directly to hurry to ftool.

Almoft every one, on their firft being feized, complained of a bitternefs in the mouth, and a continual inclination to vomit. Many brought up, juft after the cold fit, a bilious matter; fome vomited very violently the firft day, and were relieved by it; many had this propenfity to vomit, even in the progrefs of the difeafe, and continued to caft up with great benefit till the fourth day. Such as from the firft had fought for refuge in wine, and other hot things, brought up every thing they took into their ftomachs, almoft every day, complained of the heart-burn, and were in the greateft danger.

The hot fit followed immediately after the cold; and in very bad cafes, fome had the firft day an intolerable head-ach. The fever at firft appeared to be fmall in moft, but in the courfe of the diforder, was ftill more and more confiderable; yet in the moft violent fpecies, and where there was the moft danger, it was at times not obfervable, and the pulfe infinitely weak; in lefs violent kinds, the fever was often very high: I faw too, in fome, even at the firft day, a perfect delirium; in others, a continual lethar-

gy,

gy, which accompanied many in defperate cafes, and was particularly conftant in children. The difeafe was very favourable in fome, after a flight attack: thefe had little fever in the beginning, and their ftools, even the third day, continued to be yellow, and very little offenfive; but after that time, they began to complain of a bitter tafte in their mouths, and the violence of the fever increafedwith, the increafing difcolouration of their ftools.

I always found the excrements thin; but very often vifcous, and that even at the beginning of the diforder. With fome, they were quite bloody the firft day, with others, later: in thofe that were feverely attacked, as well as young children, they were mixed from the beginning,with grumous blood. I have feen children,from whom, in the firft days of the malady, the blood has flowed in ftreams down their legs; juft after appeared a quite green matter, and this gave place to a red: with moft, the excrements were at the fame time white, red, yellow, brown, green, and fometimes even black, for the moft part yielding a very bad fmell, which was at times perfectly cadaverous. The excretions in fome, who had taken no medicine, remained for a whole week quite white, and came away without pain; a week after that, red, with great pain; and throughout

feveral

feveral fucceeding weeks, red, white, and very little painful.

In flight indifpofitions, the patients went to ftool about fifteen or twenty times a day, and many forty or fifty. I faw, and even cured fome, that in the fpace of twelve hours, had from an hundred and fifty to two hundred ftools; and whofe evacuations came fo quick one upon another, that one would have thought their whole infide was coming out at their fundament.

The tormina were always more violent before going to ftool, and I thought my patients very well off, when the pains remitted after evacuation; in many they were very fharp, and in fevere illnefles drove the patient almoft to defpair. They were accompanied in the courfe of the diforder, by a fmart pain in the back, fometimes a heat of urine, and in moft perfons by a tenefmus.

In the worft fpecies, the cheft was oppreffed. In all kinds I found the appetite and natural fleep entirely gone: moft had an inextinguifhable thirft, and the greateft part were obliged to keep their beds, by reafon of their extreme feeblenefs; many were inconceivably weak, and at times fainted away. There were, however,

B 3   fome

some who were able to sit up out of bed; and many, in light indispositions, walked about. Many sweated, but without benefit.

The bad sorts of this dysentery lasted sometimes from fourteen to sixteen days, especially when proper evacuations could not be made during the first days of the disorder; though most of my patients recovered in five or six days. There appeared in some, that were very hard beset by the disorder, a rash on the mouth and tongue, in others, all over the abdomen, and in others, all over the body; though the disease, in reality, was as good as cured. In one single subject, I saw, after a happy and perfect cure, a prolapsus of the rectum. I have not experienced a relapse in any of my patients, excepting two in one person; the first proceeding from a violent fit of anger, and the other, because he got out of bed in the night, and was forced to run about the streets several times in a hard shower of rain.

They who were the most dangerously ill, had a regular miliary eruption, and at the same time, ulcers about the body, at a time when the disorder was at its greatest height, if they had not taken the purging medicine that was ordered. The greatest misfortune that attended very young children,

dren, who were very feverely feized by this malady, confifted in the fpafmodic contractions
of the nerves, which came on at its firft commencement, and by which they were immediately deprived of all fenfation.

When the difeafe terminated fatally, the tormina did not remit after going to ftool, but
were every day more and more intolerable, and
the ftools remained equal in number; a hiccough, at times a vomiting, and fwelling of
the abdomen next followed, and laftly, the cholic pains ceafed. Death brought up the rear,
(efpecially with them who had drank freely of
wine) as early as the fifth, eighth, nineth, and
fourteenth day, and fometimes later.

They that in fevere cafes, only took medicines in the beginning of the diftemper, and
afterwards laid them afide, were in very great
danger; and though they took to them again
in fix or eight days, yet ftill they continued a
long time ill, if at length they did not happen
to die. Many that took no medicines at all,
had a gentle, but tedious dyfentery; gripes,
tenefmus, and alfo blood mixed with their excretions, which otherwife had only been flimy;
great wearinefs in the members, frequent returns of the cold fit, violent fweats, indigeftion,

and

and pains in the ſtomach from every thing they eat. Others were haraſſed with a flying gout; others, among whom were likewiſe children, with a dropſy; others, again, with obſtinate ſwellings in the feet; and with others, from whom the evil ſeemed to go away of itſelf, ſtill remained a great pain in the loins, and a rheumatiſm in the joints.

The more favourable ſpecies of the dyſentery, ſhewed themſelves by an univerſal languor, a ſhivering, ſome propenſity to vomit, a cholic of no very long continuance, and much leſs frequent, as well as leſs griping ſtools. The excrements were for the moſt part white, and their food came away undigeſted; the blood did not appear till after ſome days, or elſe the ſigns of it were hardly to be perceived.

Some, in the beginning or end of the epidemy, eſpecially thoſe who lived towards the boundaries of its ravages, were only troubled with a violent griping, which continued five or ſix days, and ſometimes a fortnight, without purging, but rather a conſtipation of the belly: though when I had given them ſomething opening, I found their excretions mixed with blood and white like pus. Such as had taken no medicine

dicine in thefe circumftances, fell into a moft dreadful dyfentery.

Many had a mere griping diarrhœa, which ftaid with moft perfons but a few days, in which, however, I found the excrements frothy, and mixed with gall. A purging of this kind remained fix weeks with a lad, to whom I purpofely gave no medicines, as I hoped, that by means of this, he would be rid of a different kind of diftemper, which returned upon him every year; which accordingly happened.

Some, that were not attacked by the dyfentery, where it had raged, but had attended on thofe that were fick of this diforder, or lived in the houfe with them, at the end of the epidemy were plagued with large boils on the breaft, under the arms, on the knees, and legs; fome had them on the head, and over their whole body; many, inftead of boils, had great white bladders: yet none of thefe people kept their beds.

I have taken the whole of thefe remarks from obfervations made by myfelf, and that not with a fuperficial inattentive eye, but with all the exactnefs that the object of my profeffion required; and though, without doubt, this does

not

not comprehend every thing that other physicians might have seen during our epidemy, yet
is it quite sufficient for my purpose.

CHAP. II.

Explication of the DISORDER.

SYDENHAM's noted observation on the
close connexion between the epidemical
disorders of the same year, was fully verified
here. After an astonishing number of putrid
fevers, followed the dysentery, attended likewise with a putrid fever.

Many thousands of people in Switzerland,
and chiefly in the canton of Bern, were from the
end of the year 1764, to Easter 1765, and still
later, attacked with a putrid fever, which, for
the most part, attacked the breast, and had the
appearance of a pleurify; though sometimes the
liver was likewise ulcerated, or the intestines affected with a gangrene, after which even the heart
was found inflamed and gangrened : these last
cases, however, were very rare, and the distem-
per fixed itself mostly on the breast. In April
and May 1765, it shewed itself at the fullest
in our parts ; I also saw even in June, some sick

of

of the putrid fever at the very fame time, and
in the fame villages in which the dyfentery firft
began. In the fame manner, at the going off
of the putrid fever that ravaged with fuch violence in and about Laufanne, in the year 1765,
a dyfentery followed of the very fame complexion, which like that too became epidemic
in harveft time.

The analogy of our dyfentery, with the preceding epidemical putrid fever, appeared from
the refemblance of the fymptoms in each diforder, of the method of cure that was the moft
fuccefsful in both, and even of the effects that
followed the errors therein committed.

The putrid fever in May 1765, feized on
the children and adults in my neighbourhood,
fo very fuddenly, that they were perfectly in
health, and extremely ill, within the fpace of
a quarter of an hour; though I have for fome
days before the real attack, remarked a great
fenfation of cold in the hands and feet of fome
of my patients, together with a fhivering, but for
the moft part the affault manifefted itfelf very
fuddenly: it was juft the fame in the dyfentery.
In the putrid fever, all thofe that were violently
ill, had in the beginning an univerfal chill,
which had different degrees of duration, and
sometimes

fometimes it was long and very violent; many had only a fmall paroxyfm; with many it returned from time to time in the courfe of the fever, and went off in a hot fit: the fame things happened in the dyfentery. In the putrid fever, every one at the firft coming on of the diforder, felt a pain in all their joints, as if they had been bruifed, and efpecially in the back and loins: the fame happened in the dyfentery.

In the putrid fever, almoft every one on their firft being feized, complained of a bitternefs in the mouth, and a continual inclination to vomit. Many brought up, juft after the cold fit, a bilious matter; fome vomited very violently the firft day, and were relieved by it; many had this propenfity to vomit, even in the progrefs of the difeafe: the fame happened in the dyfentery. In the putrid fever, the hot fit followed immediately upon the cold, and very often an intolerable head-ach. The fever at firft appeared to be fmall in moft, if they did not directly heat themfelves with wine and brandy, but in the courfe of the diforder was ftill more and more confiderable; though in very defperate cafes, it had a fly, deceitful, flow manner, at the fame time the pulfe was very weak; fome that were in great danger, fell into a lethargy: the very fame happened in the dyfen-

dyfentery. In the putrid fever, the expecto-
rated matter was tinged with gall, and at times
with blood; the ftools were yellow, green,
dark-brown, of a putrid, intolerably ftinking, and fometimes cadaverous fmell : the
fame happened in the dyfentery. In the putrid fever, the excretions being mingled with
blood, was no fign at all of inflammation, as the
blood difappeared after taking the emetic : the
fame held good, with regard to the ftools in the
dyfentery. In the putrid fever, the appetite and
natural fleep were almoft entirely gone; in bad
cafes, the feeblenefs was inexpreffible the very
firft day; the patients were frequently feized with
faintings; the profufe fweat that I remarked
throughout the whole courfe of the diftemper in
thofe patients, who had neglected to take a
purge at the beginning, was of no fervice; gave
not the leaft relief, was evidently fymptomatic, and never critical: I faw the very fame thing
in the dyfentery.

   In the putrid fever, during the progrefs of an
illnefs, that continued fomething longer than
ufual, a miliary eruption broke out, and fometimes large exanthemata, that came to fuppuration, if the bilious matter was not purged off in
great quantity, directly at the beginning; fometimes this rafh was taken away by a purging,

that

that came of itfelf. In Solothurn it difappeared directly, when at the commencement of the
illnefs, recourfe was had to emetics; it is alfo,
according to the numerous obfervations I have
lately made in that city again, every where
about, fince people have endeavoured to throw
it out in all putrid fevers, during the fpace of
many weeks; which makes it palpably evident,
that the miliary eruption in fuch cafes, is very
often nothing more than an accidental effect of
a difeafe not well looked after, and therefore is
not critical, but the fruit of the method of cure.
Already, in the beginning of this century, it
has been obferved in Breflau, that a mortal miliary eruption, has fometimes appeared out likewife in the dyfentery; in that of Nimeguen, in
1736, the miliary eruption was frequently remarked at the end of the diforder; in the dyfentery that was epidemic in Zurick, 1764, miliary puftules broke out fometimes on the day
of the patient's death. In our dyfentery, I faw
a very dangerous fymptomatic miliary eruption,
in fuch, as in the beginning, had not taken the
purge according to orders; and a harmlefs critical breaking out in fuch, as had happily got
over a violent dyfentery.

In the putrid fever, the phyficians were afraid
to truft the cure of the diforder to the evacuations

tions made by nature herſelf, as art was found
to effect them much better than nature; it was
the ſame in our dyſentery. In the putrid fever,
an inflammation ſometimes joined itſelf with
other effects of putrid matter; theſe caſes were
extremely dangerous; a ſhort time too before
death, the abdomen was obſerved to ſwell, and
a gangrene on the inflamed part: it was the
ſame in the dyſentery. In the putrid fever, I
have often remarked, that the diſorder was of
long continuance, when the patient, on account
· of his weakneſs (which proceeded from that
very cauſe) would not allow the phyſician to
purge his body perfectly of its bilious matter;
and afterwards remove fully, by the uſe of proper means, the perhaps remaining corruption
of the humours; I have remarked too, that the
whole cure was unſtable and imperfect, when
they often changed their medicines, and did not
continue long enough a well indicated remedy.
The ſame I ſaw in the dyſentery.

At the time that the putrid fever reigned,
there were likewiſe many ſlight indiſpoſitions,
reſembling the other in miniature, which were
of the ſame nature, but attended with infinitely
ſlighter ſymptoms, were eaſily cured, and even
vaniſhed of themſelves without the help of medicine: the very ſame happened in the dyſentery.

tery.   The boils and large white bladders, that were obferved towards the end of the epidemy, in thofe that remained free from the dyfentery, were befides well worth notice; thefe, however, fhewed alfo a corruption of the humours.

In the putrid fever, all depended on the fpeedy evacuation of the bilious matter.   The fever kept off, as foon as the corrupt matter was for certainty purged out of the body; fo that after this manner, the cure of this otherwife dreadful diforder, appeared to me not in the leaft difficult, for I have cured a great many putrid fevers, in two, three, four, five, and fix days : a ftrong proof that a good method of cure is the beft fpecific.   I made no fcruple of procuring farther evacuations, when the danger appeared very great, and other phyficians would have abfolutely given the patient over to his fate; for in the greateft weakneffes, I gave emetics with the beft effects, on the eleventh, and even after the twentieth day.   I have alfo taken away bad confequences of the putrid fever, by purgatives; for example, an extraordinary obftinate cough, with the fulphur auratum antimonii : all this happened with the fame fuccefs in the dyfentery.

In

In the putrid fever, I reckoned among my capital remedies, ipecacuanha, tamarinds, cream of tartar, all vegetable acids, and the fulphur auratum antimonii : in the dyfentery, thefe fame remedies had the beft effects, and inftead of ful-phur auratum antimonii, I made ufe of the vi-trum ceratum, with the greateft fuccefs.

In the putrid fever, I was under the greateft apprehenfions, when every thing inclined for the beft, as at that time the patient, or the afliftants, very eafily committed an error in the non-natu-rals, that became mortal : I feared the very fame thing, for like caufes, and with equal rea-fon, in the dyfentery. For in the putrid fever, I found wine, juft as in the dyfentery, very hurt-ful and dangerous.

But I difcovered the moft ftriking refemblance between our putrid fever and dyfentery, in the manner, in which, after the ufe of the before-mentioned remedies, the fymptoms of the putrid fever vanifhed by degrees; and on the neglect of thefe, or ufe of medicines of a contrary na-ture, became obftinate, and again got the upper hand : in fhort, in the manner in which they appeared under different fhapes, and degenerat-ed into the moft dreadful fymptoms, when the

C   patient

patient did not do his duty, fo well as the phyfician.

From this perfect refemblance of the putrid
fever with our dyfentery, we can therefore moft
clearly determine the fpecies of this laft mentioned diforder. The reader fees, without my
reminding him, that our dyfentery was accompanied with a bilious, or, as it is called,
putrid fever.

As the putrid fever was only catching in certain circumftances, fo our dyfentery was of itfelf
not contagious. I have feen a great many people keep company with the fick, without any
detriment to themfelves ; but many followed at
the fame time the advice I gave them for avoiding
the infection ; feveral did not follow it, and yet
were not infected. In many houfes almoft every
body was ill, and that not at once, but one after
another ; in many, I faw only one perfon fick.
I do not indeed wonder, that one fhould be infected, and another not, as the conftitution of
the body, and even of the mind, can make one
man much more fubject to infection than another. The power of contagion in the dyfentery,
is alfo very different; while in an epidemical dyfentery, that is, otherwife attended with a putrid
fever, the diftemper may differ very much, according

cording to the degree of the putrefaction; all thofe that were dangeroufly ill of our dyfentery, had the putrid fever in the higheft degree; on the contrary, in fome gentle cafes, many had not the leaft fymptom of it, nor were their ftools in fo great a degree offenfive. Now the contagious power of the dyfentery, lies chiefly in the excrements; for the mere fmell of them, has often communicated the dyfentery to men in perfect health, and even beafts; Dr. Pringle even faw one that proceeded from the bare fmelling to blood that was putrified, by being kept in ftopt bottles; as in general the effluvia from putrid blood, are more apt to caufe a dyfentery, than any other malady. And even though one ftop one's nofe, one is not fecure from infection; for the putrid vapours adhere to one's cloaths, and when they are in a high degree contagious, are thus communicated from one perfon to another; while, at the fame time, he that has the cloaths on his back, is perhaps not in the leaft infected. The fœtus was naturally infected in our dyfentery, when the mother lay fick of it herfelf: a woman in the city of Frawenfield, that was troubled with this diforder, a fortnight before and after fhe was brought to bed, brought her child likewife into the world infected with the fame, and it died three days after: but this cafe proves nothing at all. In general, it ap-

peared

peared to me, that our dyfentery became contagious purely through naftinefs, and the croud-
ing many people together in a fmall fpace, but
was by no means fo of itfelf; for though many
were attacked with it at once, this feems to proceed from a more univerfal, and widely different
caufe, which operated at once on every one.

After this exact determination of the fpecies,
under which our diftemper fhould be ranged,
I betake myfelf, with the greateft fear and caution, to enquire into its remote and proximate
caufes, as they are called. People of narrow
capacities, who think that learning confifts in
knowing every thing, will be very ill content
with this enquiry. I on the contrary, muft here in
many things, as on moft fubjects in every thing,
confefs my ignorance; fince it is much wifer, as
well as better, to obferve narrowly the works of
nature, than to explain their caufes by arbi-
trary hypothefes.

The weather this year feemed perfectly favourable to this dyfentery. The air was in June
very inconftant, but for the moft part humid;
and when the fun broke forth, there enfued a
fuffocating heat: July was full as changeable;
but the heat never rofe to fo high a pitch: Auguft was, during the better half, cloudy and
rainy;

rainy; afterwards the days were fine and very
warm, and at the fame time the nights extremely cold: till the middle of September, and longer, the fky was continually clear; at noon, it
was extraordinary hot, in the morning, evening, and chiefly throughout the whole night,
intolerably cold, afterward the air grew foggy,
damp, and cool; and we had fine weather and rain
by turns: October was very variable, though for
the moft part cool; and the clofe of the month
was ftormy, rainy, and pretty cold. By means of
thefe great changes from heat to cold, the perfpiration was by turns firft promoted, and afterwards
on that very account the more violently checked;
thus the putrid fcum of our bodies remained for
the moft part behind, and was forced to empty
itfelf into the inner cavities. I have, indeed, remarked, that thofe chiefly were taken with the
dyfentery, who after having very much heated
themfelves, cooled themfelves directly after; efpecially fuch as drank great quantities of cold
water, when they were in a profufe fweat. This
feemed to be the reafon, that moft of our peafants fell fick of the dyfentery.

In general, it is not the cold that follows
on heat, and remains, but that which fucceeds
heat, and gives place to it by fits, that is confidered as the caufe of the dyfentery. The cold

air

air before fun-rife, which is followed by a
fcorching heat at noon, after which the nights are
again cold and damp, are thought to be the
principal occafions of the malignant difeafes of
the army in Hungary; and in particular, the reafon why autumnal fevers and dyfenteries are
more frequent and violent in that country than
elfewhere. Now we had, this year, fuch weather for the moft part, where the dyfentery was
moft violent; but in a great many places, in the
fame weather, and at the fame time, it was not
remarked; it came on alfo, when this change
from heat to cold was not at all obferved. Sometimes it made its appearance, when in the beginning of the fpring, a fudden heat enfued after a great cold; and on the contrary, the paf-
fengers in the Dutch fhips, faw both the diarrhœa and the dyfentery increafe, in proportion
to the coldnefs of the countries they paffed
through; though otherwife, the epidemy of the
dyfentery was moftly ftopt by the approach of
the cold. The hippocratic (if I may be allowed
the phrafe) foutherly winter of 1764, appeared
to us to have occafioned the great quantity of
putrid fevers with which we were plagued; but
other winters, to which it bore a perfect refemblance, did not bring thefe fevers on, and we
have putrid fevers even in the coldeft feafons;
for the extraordinary cold winter, in the beginning

ning of the year 1766, was exactly the time in
which the putrid fever, especially the putrid
pleurify, and even the malignant fever, raged
more violently in Switzerland than ever they had
done before. Generally speaking, the same diforders do not always show themselves in the
same weather; and diforders, that perfectly re-
femble one another, often make their appearance in totally different kinds of weather. I cannot therefore comprehend, why fome people explain the manner and ways, in which a certain
particular ftate of the air has given rife to a
particular epidemy, with as much confidence, as
if it could not possibly be otherwife. All that
I can therefore conclude with any precifion, from
the before-mentioned obfervations, is this; that
cold alternating with heat, certainly contributed
very much to our dyfentery.

But the reafon why this malady did not, in
like weather, break out in fo many other places,
is to me unknown. Without doubt or contradiction, this year's dyfentery arofe from a corruption of the humours; as an exact obfervation of the diforder, taught me in the cleareft
manner. It is full as obvious, that a certain
concourfe of caufes, before-hand internally exift-
ing in the human body, is requifite to produce
a difeafe, that neverthelefs attacks it fuddenly;

for

for without fuch a concourfe of internal caufes, every one would certainly have had the reigning malady, and in the fame degree. This union illuftrates at once many undetermined, and in part, contradictory notions; and the confideration thereof, appears to me one of the principal objects of our art: where this exifts, a man is attacked, and where it is not, remains free. I have remarked, that where the putrid fever does not reign, thofe that are fubject to much vexation, and the excretion of bile that proceeds therefrom, are feized, rather than others, with that diforder. The moft inconfiderable external caufes are capable of producing the greateft effects, when joined with internal, already prefent.

It is allowed, by almoft all the moft learned phyficians in Europe, that the dyfentery (that is, with their leave, the dyfentery attended with a putrid fever) is brought on by thofe caufes, that make our juices very putrid, and incline their courfe principally towards the inteftines.

The camp dyfentery arifes, for the moft part, from a repulfion of the perfpirable matter; when the foldiers lie in the field in all kinds of weather, and muft do their duty, at the fame time that their humours are grown thin and fharp,
through

through the intenfenefs of the heat. In general, it fhows itfelf juft after the armies begin to appear in the field, grows often frequent and confiderable fo foon as the end of June, ftill more fo toward the end of July, and thus remains till the troops go into winter quarters. In the night after the battle of Dettingen, the 27th of June 1743, the Englifh foldiers lay on the field of battle without tents, expofed to a heavy rain; and the next night likewife, after marching to Hanau, lay again in the open field, on wet boards, without ftraw. The fummer had begun, and the heat had been hitherto great and continual; but the free and uninterrupted perfpiration, that was the confequence of this continued heat, had till then prevented the rife of any epidemic malady. But now the pores of the fkin were fuddenly clofed, the humours tending to putrefaction were tumid in the bowels, and occafioned a general dyfentery, which lafted a great part of the fummer, and in a few weeks had feized almoft half the army. Such of the officers (among whom, indeed, it was not fo general) as had lain wet at Dettingen, were attacked by it firft, the reft received it by contagion; but a regiment that had not lain in the damp, nor been expofed to the rain, remained perfectly free from it, at a fmall diftance from the camp, though (excepting that they were not fubject to

the

the contagious effluvia of the reft) they breathed the fame air, eat the fame provifions, and drank the fame water.

According to Dr. Pringle's obfervations, the dyfentery rages without interruption, in camps that are perfectly dry and airy, after a continued and violent heat. But in fuch a camp, befides the natural damp of a tent, thefe people muft be often expofed to wet floorings and cloaths, and to the chill and damp air of the night. The occafion of thefe diforders is the more inevitable, as in the field the variations from heat to cold, are much more fenfible and frequent than in quarters.

The dyfentery is feen in all parts, where after much heat, the perfpiration is ftopped, not only by means of a wet floor, or mifts and dews, but alfo of damp cloaths. In the warmeft countries, the dyfentery reigns when the weather is rainy ; and in all climates that are fubject to conftant rains, this diforder fhows itfelf very often. It returns alfo every time one catches cold, after a feeming, and often, a perfect cure. Baron Van Swieten thinks, with the greateft reafon, that cold feizing the body when it is heated, has killed more than the plague.

With

With thefe external caufes, internal may alfo concur. Thefe compofe the feminium of the diforder, and may lay awhile lurking in the body, till they break out on the perfon's catching cold. In fummer, not only the folids are relaxed, but alfo the juices, through the intenfenefs of the heat, incline to putrefaction ; now, when a fudden ftoppage of the perfpiration fupervenes on relaxed fibres, and a putrid ftate of the blood, it is no wonder that a dyfentery fhould be the confequence of this acrimony, formed within the body. Of all our humours, the gall is principally fubject to putrefaction. Hippocrates, indeed, afcribes the difeafes of fummer and autumn proceeding from this caufe, to a redundancy of bile; but moft other writers, to its putridity; fo that as well of old, as fince, thefe diforders have been in general called bilious. Now we learn from a number of obfervations, that in the bodies of fuch as have died in the dyfentery, the bile was faulty, as well in quantity as quality; that either none was there, or that which was there, was perfectly acrid and corrofive. We may therefore take it for granted, that the bile in fummer time, if not more abundant, is yet at that time more corrupted than ufual; and that this circumftance, if it be not the primary caufe, is, at leaft, the confequence of all fummer and autumnal diforders, and increafes

creafes their malignity. In a very general and epidemic dyfentery, an acrid and corrofive bile has commonly the moft confiderable fhare; fo that our dyfentery feemed to me, nothing more than a particular determination of the very fame corrupted humours, which occafion our putrid, or rather bilious fever.

From all this enfues, that the ftate of the air during this year, conduced very much to our dyfentery; that likewife this dyfentery proceeded from a putrefaction of the juices, which in particular cafes, is very eafy to explain; the general caufe of which however, I, confcious of my own ignorance, leave to others to determine.

The proximate caufe of our dyfentery, the manner and wife in which it fhowed itfelf in the body, and its effects, can now be declared with greater eafe and probability. I faw, with my own eyes, that a corrupt, putrid, and bilious matter, lay in the ftomach and inteftines, caufed great pains, and at firft upwards, and afterwards downwards, fought an exit out of the body. Now it is well known, that the bile, from various caufes, may be fo altered, and put on fuch a fharp, putrid, and acrid nature, as, like a poifon, to corrupt the whole body; for from this corruption, either inflammations, ulcers, and gangrenes,

grenes, take their rife, or all the juices in general fall into the like putrefaction in divers degrees, and produce miliary, or petechial eruptions; fometimes it happens, that the gall of
itfelf affects this fharp, corrofive, and perfectly
poifonous nature, and thus infects the other juices
of the body. Sometimes this is preceded by an
acrimony, formed in the blood, whether it proceed from contagious vapours, (as happens when
many fick perfons are crouded together in our
military hofpitals) or from what other caufe foever, and this corruption is communicated to
the gall. The remark, that the dyfentery has
fometimes owed its origin to the bare fmelling
to blood, that was putrified by being kept in a
ftopt bottle, makes good, thus far, that expreffion of Sydenham's, who calls the dyfentery a
fever that has thrown itfelf on the inteftines.

Now when once fuch an extraordinary acrid
matter is in the bowels, it is very intelligible,
that thefe extremely fenfible parts muft be very
much irritated. This irritation communicates
itfelf to the ftomach, which gives rife to vomitings; in the bowels each irritation provokes a
greater conflux of the moft liquid parts of the
blood through the inteftinal glands into the inteftines, whence arifes a purging. Now this conflux may, without doubt, be exceedingly great,

as

as we know from anatomy, that befides the
larger glands, the liver, and the gall-bladder,
there are an infinite number of fmall paffages
over the whole furface of the bowels, through
which even the moft unufual things can pafs out
of the blood into the bowels, and that a continued and ftrong irritation of thefe innumerable
paffages, occafions an incredible flux of the humours into the bowels; and this again fuch an
incomprehenfible purging, that at fight of the
evacuations, one is apt, not very improperly to
fay, that their whole infide is coming out at
their fundament.   In this manner, in our dyfentery, above forty pounds of a watery matter has
been feen, in one day, to come away by ftool.

From thefe phænomena, we are enabled to explain the reafons of the dreadful pains in the belly, and other fymptoms. The bile does not always
occafion the pain; as there are dyfenteries, in
which the patients void not the leaft bilious matter; and, as in the putrid fever, very feldom a pain
arifes, from the mere prefence of this matter in
the inteftines.   But yet the belly-ach in the
dyfentery is very often at firft a confequence
of the irritation of the acrid and putrid humours
in the bowels, and of the fpafmodic contractions
in thofe parts, produced by that irritation: in
the progrefs of the diforder, it is the confequence
of

of the abfence of their natural mucus, which makes the bowels, deprived of this mucus, more and more fenfible to each frefh irritation; fo that the greater acrimony of this matter, and its being accompanied with a fever, diftinguifh the dyfentery from the diarrhœa. The tenefmus is the effect of an irritation in the inteftinum rectum, the prolapfus ani of the tenefmus, and the ftranguary of an irritation in the neighbouring parts.

But the excrements in the dyfentery, of which we treat at prefent, do not merely confift of corrupted gall; befides, all that a man voids of a green or yellow colour, is not pure gall, fince one drop of bile colours an aftonifhing quantity of water. They are very often white, and perfectly refembling pus; though it is for the moft part a great error to take this matter for pus. For it is well known, that the inteftinal glands, in the fame manner as thofe of the urine-bladder, when torn with the ftone or gravel, are capable of yielding a greater quantity of fluids, and thofe of a quite different nature than in health; this humour is in both cafes a flimy white matter. On comparing thefe remarks with the foregoing, it is manifeft, that ftools of this kind may be produced, by an acrid bilious matter adhering to the bowels; and that, as my obfer-

vations

vations in the firſt chapter ſufficiently ſhew, in
a bilious dyſentery attended with a putrid fever
the excrements may be even perfectly white.
From this appears, by the bye, how ridiculouſly
ſome people conſtitute different ſpecies of the
dyſentery, from the different colours of the excrements, and treat them by quite different
methods.

The particles of fibres and membranes, that
often come away in the dyſentery, hanging ſometimes a foot long from the poor patients, and
are conſidered as the inner coat of the inteſtines,
are in reality very ſeldom any part of them, but
often nothing elſe than an inſpiſſated mucus.
Great anatomiſts have demonſtrated to us the
paſſages, through which this mucus comes into
the inteſtines; and withal, that a ſubſtance can
come out of the blood into them, by which this
mucus is coagulated, and under the appearance
of a fleſhy, membranous, or fat body, paſſes
into the ſtool, when, at the ſame time, not the
leaſt ulcer in the bowels is to be perceived. I
do not, however, deny, that the tunica villoſa of
the inteſtines, is not alſo ſometimes abraded, and
comes away with the excrements. I am likewiſe ſenſible, that the bowels are in this diſorder
apt to be ulcerated; but ſo late, that this matter is changed into a putrid thin pus, or is ſo

con-

confounded with blood and mucus, that one cannot poffibly fee it. Hence we may perceive, how often phyficians deceive themfelves and others; when they, in the very firft days of the diforder, miftake the forementioned mucus for pus, the fibrous and membranous fubftances for figns of the laceration of the internal tunic of the inteftines, or of an ulcer in thofe parts; and thus in a bilious dyfentery attended with a putrid fever, entirely omit purgatives, and give the patient over to death.

The bilious, putrid, and corrofive matter, inclofed, as it were, in the cavities of the inteftines, irritates them fo much; that often, the openings of the blood-veffels into the inteftines are widened; fo that pure blood runs out, and mixes itfelf with the ftools. Thus there may be blood in the excrements, without the leaft perception, or even fufpicion of inflammation in the bowels; it may alfo flow in great quantities, without their fuppuration enfuing: hence appears the reafon, why, when even the excretions are bloody, there is no need to be afraid of expelling the bilious irritating matter with a vomit and purges, and why it fo often arrives, that a vomit alone puts a ftop to this flux of blood. Nor is an internal heat, which the egregious Mr. Rahn, in his work on the dyfentery,

D   affirms

affirms to be an infallible mark of a violent inflammation of the bowels, any more a sign thereof, than bloody stools ; for I have removed this symptom likewise by means of tamarinds, which evacuated the corrosive bilious matter; while, in case of the slightest inflammation, this ardor had been violently increased.

Notwithstanding what we have said, a bilious dysentery (or the dysentery attended with a putrid fever) may turn to an inflammation of the intestines and a gangrene, in the same manner as a putrid fever very often ends in a gangrene of those parts. Stools perfectly black, and of a cadaverous smell, cold sweats, the hiccough, and delirium, are considered as tokens of a gangrene in the bowels; and perhaps it is not easy to find an epidemical dysentery, in which, towards the end, the bowels are not inflamed. Almost in all parts of the primæ viæ, inflammations, suppurations, a number of ulcers and gangrenes are found after the dysentery; but in every one, these evils are conspicuous in the rectum and inteftina craffa, which, for the most part, are putrid and mortified. Sometimes small pustules have been seen in the bowels, which came away even during the patient's life, and were full of a putrid stinking matter; and in the inteftina craffa, little aphthæ, that bled when

preffed,

prefled, and looked like the flat kind of fmall-
pox, when this diforder is at the higheft, but
with this difference, that they were folid and
without any cavities; they confifted of the two
innermoft tunics of the inteftines, that grew one
within the other, and were incraffated by the
inflammation: the firft of thefe was covered
with a black mucus, and black fpots were likewife partly vifible on it; fometimes the mefente-
ric glands are fwelled, relaxed, filled with a bad
kind of pus, and very nearly mortified.   Nay,
after tedious dyfenteries of long ftanding, inflammations are often found in the rectum, in the
reft of the inteftina craffa, fometimes in the fmall
inteftines, and even in the ftomach. ·

   The tenefmus, however, ought very feldom to
be confidered, as a fign of an inflammation of the
inner coat in the extremity of the rectum, or of
an ulcer in thofe parts.   Great anatomifts are of
opinion, that what ftimulates the rectum to evacuation towards the end of a dyfentery, is no particular affection of the gut itfelf, but the remainder of a fharp humour, and fometimes of the
blood itfelf, (when the excrements are of a dark-
red colour;) in the next place, that this refiduum
may ftay awhile in the cavity of the great guts,
and by degrees be pufhed into the rectum and fo
towards its extremities, which are, at that time,

very fenfible to this irritation. Yet, it is not without reafon, that from a violent tenefmus remaining after the dyfentery, fome people have fufpected an exulceration of the rectum, or fome other confiderable complaint; as the fequel has fhewn this affertion to be true. A tenefmus has been really known to proceed from an inflammation of the rectum, that lafted feveral days, fometimes a week, and longer, and was afterwards followed by a greater or lefs excretion of a yellow pus, upon which the tenefmus ceafed.

This effay towards an explication of our diforder, with all its imperfections, may not perhaps appear totally ufelefs, when the influence of fuch refearches on the cure is confidered; and thence is perceived, how fenfelefsly phyficians of an ordinary caft, on occafion of fuch enquiries, pour out a volley of abufe againft theory.

C H A P. III.

The Curative Indications, Diet, and Prophylactic Remedies.

THE foregoing explication of our diforder, difcovers, at the fame time, the indications towards its cure. Firft of all, an enemy of itfelf fo dangerous, and which would

become

become ftill more fo by a longer ftay in the
body, fhould be quickly expelled from it, and
the putrefaction, at the fame time, be refifted in
the beft manner poffible.

In no malady does nature fooner get the ftart
of art than in this, if one does not in the very
beginning take care to do, what is often afterwards too late to be done. The morbific matter of a dyfentery of this kind, has been likened
to a rotten egg, that given to the quantity of a few
grains, occafions a violent and continual vomiting: the putrid ftagnating gall is not lefs venomous, for which reafon, it fhould be without
delay driven out of the body. White and red
miliary puftules, and even petechiæ, are, in putrid fevers, an effect of the putrid matter that
has paffed into the blood; the fame evils are
produced from like caufes in the dyfentery.
Not to evacuate, or correct immediately this
matter, is to produce the miliary eruption that
is fo often mortal; but thefe excrements never
break out in the bilious dyfentery, and feldom
in our putrid fevers, when evacuations are procured in proper time and quantity. Add to
this, that the greateft phyficians allow it to be
moft commonly difficult, and often quite impoffible, to cure a dyfentery of fome weeks continuance.

It

We muſt therefore, not only abſtain from the pernicious practice of ſtilling the diforder in the beginning, before we look out for more effectual remedies; but we ſhould alſo take care above all things, not to ſtumble on ſuch remedies as may retain in the body, an enemy ſo extremely dangerous. In ſevere caſes of the dyſentery, the forces of nature never ſuffice alone to expel the putrid matter out of the body, though nature endeavours to do it by this very way : and in all caſes, it is contrary to her intention to retain it. The obſervation, that inflammations are apt to ariſe, in diſeaſes of a putrid nature, and that in moſt of thoſe, who die of the dyſentery, the bowels are inflamed, indicates in like manner, that every thing ſhould be avoided in this diſtemper, which immediately produces fevers and inflammation.

Theſe indications towards the evacuation of the putrid matter I purſued with emetics, given directly at the beginning, when nothing particular hindered. Nature herſelf·pointed out this way, as almoſt all that were taken with the dyſentery, had, at the beginning, a continued propenſity to vomit: many vomited, and that with benefit. I choſe gentle medicines, as with theſe we can effect as much as one can wiſh : I gave theſe emetics, even at a time when the ſtools were very bloody; as I ſaw that leſs blood came away af-
ter-

terwards: I even gave a vomit very late, when I
was called late, and no evacuations had been
previoufly made. I always omitted the vomit,
when there was the leaft fufpicion of inflammation, or when circumftances of a quite different
nature forbad it; as, for example, a rupture; or
likewife in cafes of very young children, by reafon of the parents fears; for which, however, in
other maladies of fuch infants, I have fince fhew-.
ed lefs indulgence; and particularly in the hoop-
ing cough, in 1766, with the greateft fuccefs.

After the emetic, I purfued thefe fame indications with purgatives. Thefe were indifpenfibly neceffary, as the acrid putrifying matter,
that lay burrowing in the inteftines, from the
prolongation of its ftay in the bowels, grew
more and more acrimonious, irritating, and apt
to fpread the putrefaction in the body, or to
caufe inflammations in the inteftines. The
bloody ftools did not hinder me from giving
purges, as I faw that they cleared the bowels of
the acrid matter and that no more blood appeared in the ftools, as foon as this matter was
quite evacuated. I gave thefe purges, as long
as there were any indications of an acrid putrid
matter in the bowels, without inflammation or
fuppuration. But thefe purges were always of
a gentle nature, and, to chufe, of the acid kind;

as

as ftrong purgatives in the dyfentery always occafion an intolerable cholic, and greatly weaken
the patient; and cathartics of an acid nature do
not only for the moft part evacuate the putrid
matter, but alfo withftand any putrefaction that
may ftill remain in the body.

· This was the chief thing to be attended to,
after, and even during the evacuation. The
putrid venom fhould be oppofed, and the humours of the body preferved from the like corruption. Quite above the prejudices of the vulgar, and indeed, of great phyficians, particularly
thofe of Dr. Degner, I gave with this view immediately in the beginning acid falts, not only in
fmall, but alfo in very large dofes. I followed
likewife this plan in the diet I prefcribed to my
patients.

The pain could not be more effectually relieved,
than by taking away the fharp corrofive matter.
Yet was this fometimes fo very abundant, that in
fpite of all evacuations, enough remained behind to keep up the diforder, and excite the
moft intolerable pains. In thefe cafes I very
feldom made ufe of anodynes of the narcotic
kind, and never without the greateft caution
and referve. But in thefe and all other circumftances, I was extremely attentive to affift the
bowels

bowels with a proper quantity of foft and mucilaginous liquors; fince except in the cafe of inflammations, thefe pains are folely to be afcribed
to the prefence of an extremely acrid matter in
the inteftines, which irritates them to fpafmodic
contractions, and to the lofs of their natural
mucus in the courfe of the diforder.

The methods I took to fortify the ftomach
and bowels, after the dyfentery, confifted in
making choice fometimes of fuch remedies,
as, while they ftrengthened, promoted fome
degree of evacuation, and at other times of fuch
as ftrengthened without heating. For the moft
part, I had no indications at all of this fort to
purfue, as my patients recovered their ftrength
of themfelves.

The diet was in general regulated by the
caufe of the diforder, and the peculiar circumftances of the patient. With regard to the air,
I was extremely careful to have it always frefh
in the chamber; though I gave them great
caution with refpect to taking cold, which is of
fo much prejudice in this malady. With regard
to victuals and drink, I took care, above all
things, to avoid every thing that promoted putrefaction; though at the fame time, I enquired
narrowly into the nature of thofe kinds of food,

that

that are reckoned to promote or refift putrefaction; for, though Degner had in the dyfentery at Nimeguen, which was of the putrid,
(as it is called) or bilious fpecies, the fame
indications to oppofe putrefaction, yet he allowed his patients different kinds of flefh-broths,
which, however, promote it, and eggs, which
are manifeftly of a putrid nature. Our excellent countryman, indeed, Mr. Conrad Rahn,
advifes chicken and veal in the dyfentery, as he
imagines that they expel wind. But this entire
new difcovery was at that time unknown to me;
befides, I do not fee why one fhould think on
carminatives; and once for all, any flefh of what
kind foever, is hurtful in all putrid and inflammatory fevers; as in the former they increafe the
putrid corruption of the humours, and in the latter, ftill more incraffate the blood already of itfelf
too thick and vifcid: accordingly I forbad all
flefh and flefh-broths, as well as eggs, which
are fo much recommended by all phyficians.
As for Mr. Rahn's indications towards expelling the wind, I did not follow them at all; for
the wind in the dyfentery (which however I
have feldom obferved) is the immediate effect
of putrefaction: as, according to Dr. Pringle's
experiments, putrefied animal fubftances do not
only of themfelves produce air, but likewife excite a violent fermentation in all vegetable
foods,

foods, and I could not comprehend how putrefaction could be hindered by veal, when all flesh so evidently promotes it. On the other hand, I found that carminiatives were esteemed in the dysentery, on account of their having been found serviceable in pains of the bowels, proceeding merely from wind; and hence it was very absurdly concluded, that they would be of like service in tormina that proceed from quite different causes. I therefore forbad cummin seeds, and cummin sauce, as likewise the drink the Italians are so fond of, made of a decoction of coriander seeds; but in particular all hard and indigestible food. With a general view to hinder putrefaction, I forbad all fat, butter and oil. On the contrary, I ordered to all my patients barley-water and rice-gruel, and with each I mixed, for the most part, cream of tartar; after the purge, I gave a preparation of barley, which is nothing else than barley-water boiled down to a thick and strong consistence, and strained through a cloth: this served the patients instead of victuals, and when they chose, for drink.

With regard to washing away, and attenuating the bilious acrimony, I considered copious drinking as very serviceable. Indeed, some old women of the last century, were of opinion, that a great

part

part of the cure in the dyfentery, confifted in
abftaining from drinking, which, in their opinion, increafed the purging; and that thofe were
the eafieft cured, who could fuffer thirft with
the greateft patience.   But the experience
of latter years has taught us, and very great
phyficians, Baglivi, Huxham, and Tiffot, have
advanced, that copious drinking is no where
more neceffary than in the dyfentery.   Water,
once fo much defpifed, is, when drank in great
quantity, an univerfal remedy in this malady, .
bilious diforders, and ardent fevers.   In the
cholera morbus, as it is called, or continual vomiting, Dr. Degner himfelf drank, in the fpace
of twenty-four hours, four and twenty pints of
warm water; and again in fourteen hours, forty-
eight pints more; and again in two hours, near
thirty pints, with the greateft benefit.   To
be fure one has need of an Herculean ftomach,
to bear a deluge of this kind.   In the mean
while, thus much is certain, that the drinking
of warm water in the dyfentery, is of great fervice; and that a vaft quantity of dyfenteries were
cured, only by taking a tea-cup full of warm
water every quarter of an hour.   In like
manner I allowed whey in great plenty, and rather preferred it to water, and my patients
found it agreed with them very well.   Cold liquor was, in the beginning of the diforder, al-

ways

ways noxious; on the contrary, warm drink not only does no harm, but alfo wafhes out the inteftines better, and paffes eafier through the mefenteries and lacteals.

Every thing that was capable of binding or heating, I forbad.   I difapproved of milk, cream, oatmeal, rice, and acorn-gruel; inftead of oil, I made ufe fometimes of an almond emulfion, and a folution of gum arabic.   I utterly condemned all kinds of paftry, cheefe, fpice, fpirituous liquors, and efpecially wine.   I gave alfo ftrict orders againft holding the ftools in the body, as a very pernicious practice: though thefe various orders and prohibitions, appeared in the eyes of our people, an extremely ftupid and damnable herefy.

I confidered cleanlinefs as a thing of the utmoft importance in the cure of the dyfentery. I took care of this in all refpects, and advifed, that the childrens clouts fhould be carefully wafhed.

Thofe that were getting well, I allowed as much boiled fruit as they chofe, with lemons and lemon-juice, together with the gruels before-mentioned; or alfo, when they were further advanced, a light food, compofed of almonds,

monds, milk, the white of an egg, and fugar.
After fevere illneffes, I advifed them to regulate
themfelves for fome weeks, juft as if they in
reality ftill had the dyfentery; and I repeated to
them over and over enough to make them fick,
that errors in diet, and efpecially an obftruction
of the perfpiration, from expofure to a moift
air, for the moft part caufe a relapfe, and the
fecond illnefs is always more dangerous or more
tedious than the firft.

According to the obfervation of Dr. Moehrlin, of Ravenfburgh in Swabia, a ftrict diet was
not only of the greateft fervice in curing the dyfentery; but alfo after a full remiffion of the
purging, for at leaft a week or longer, (notwithftanding a return of appetite) the ftomach
fhould not be burthened with food; as otherwife the cure would be retarded, or a relapfe
enfue. Afterwards a milk diet agreed very well
with his Swabifh patients, but flefh meats very
ill, for fome weeks after; and even good ftrong-
bodied wine, was the laft thing that Dr. Moehrlin found his convalefcent patients could bear.

The prefervative remedies were inftituted according to the moft accurate obfervations and
experience.

At

At the time, when uncommonly cold nights came on after very fultry days, I advifed my acquaintance not to heat themfelves too much in the day-time, and either not to go out at all after fupper, or if they fhould, to cloath them-felves very warm. I myfelf in the month of September, on account of the great heat of the day, was ready to faint away, when I arrived at the chambers of fuch of my patients as lived at any diftance; accordingly in the day-time, I wore the thinneft and lighteft cloaths; on the con-trary, I was obliged in my night vifits, to wrap myfelf up clofely in the thickeft cloth. I or-dered the peafants in particular, not to lay them-felves down and take their naps, according to their laudable eftablifhed cuftom, on damp floors.

Experience has fhewn, that the fmell of the patients was leaft dangerous, that their breath was worfe, and their ftools worft of all; that a fhivering is commonly the firft effect of the infec-tion, and that a vomit is of fervice in fuch cir-cumftances. In the, chambers of the fick, I generally kept a window open all day long; or ordered them to be aired twice a day, by opening the door and windows, with the bed-curtains drawn; and the air to be frequently purified befides, by pouring vinegar on a hot iron fire-fhovel. In the villages, I ordered the
excrements

excrements to be taken out of the houfe as often as poffible, and thrown into deep pits, digged for this purpofe in the meadows, at a proper diftance, and every time covered with frefh earth. In the interim, I bad them cover the ftools up clofe that remained in the houfe; and very ftrictly forbad the peafants to throw them on the dunghill, or into the ftreets. I prohibited thofe that were in health from lying with the fick, or eafing themfelves on the fame clofe-ftools, as the fick ufed. I took care that thefe laft fhould often change their linen; I commanded likewife, that they fhould take a precaution here very neceffary, namely, not to keep the dead bodies too long in the houfe, or at leaft to carry them to a feparate and cool place. It is of importance alfo, to interr the corpfes in deep graves.

As a very good prefervative, I advifed thofe that were in health to eat lefs flefh meat; and on the contrary, as much fruit and grapes as they chofe; as to the reft, to keep to the moft digeftible food, and of all things to drink wine; as it is a prefervative againft fear, and makes the ufe of the cooling prophylactic remedies fupportable even by weakly perfons; whence it limits the application of the maxim, that every thing that generates wind and relaxes, inclines the body to putridity.

putridity. I could not too often repeat to the
peafants, that after having heated themfelves,
they fhould not pour warm water too foon, and
in too great quantities, into their bodies. Dr.
Moerhlin remarked in Ravenfburg, that thofe
were either not attacked by the dyfentery, or
elfe had it very flightly, who eat little, drank
ftill lefs, and that not cold; and kept their bodies in a ftrong perfpiration, as well in the day-
time, as efpecially in the night, carefully covering
themfelves all over with the bed-cloaths.

I gave the nurfes, when they were in imminent danger of being infected, a vomit, with the
beft fuccefs; to others tincture of rhubarb, and
to moft that afked for it, cream of tartar. I was
myfelf, from too great fatigue and vexation, in the
beginning of the epidemy feized with a violent
pain in the belly, and an evacuation of an atribilanious fpumous matter: this indifpoficion returned
a fecond time, and went away each time as faft as
it came; my only prefervative confifted in ftrong
dofes of cream of tartar, and in a certain indifference with regard to the diftemper, whenfoever
it fhould attack me. In general, in this as in all
other epidemic diforders, one of the beft, though
(to put in practice) one of the moft difficult
prophylactics, is, not to be afraid. For fear
is more pernicious, than the very worft condi-

E   tion

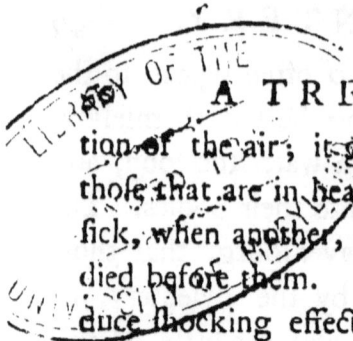

tion of the air; it gives the reigning diſeaſe to thoſe that are in health, and very often kills the ſick, when another, whom they have loved, has died before them. This paſſion and grief produce ſhocking effects on the healthy, and infinitely worſe on the ſick.

CHAP. IV.

General and particular Methods of Cure, and their Effects.

THE principal indication in each patient, was the quick evacuation of the corrupt bilious matter.

In the beginning, I effected this by a vomit, that conſiſted at moſt of forty grains of ipecacuanha, and the weight was diminiſhed in proportion to the age and other circumſtances of the patient. I ordered it to be taken in a ſpoonful of warm water, or weak camomile tea, drinking two tea-cups full of the ſame immediately after, and as often as the vomit enſued, repeating the like quantity of the drink.

Stronger emetics than theſe, I did not find fit for my purpoſe; with milder I ſhould have done nothing at all; beſides it is well known, that ipecacuanha does not relax the ſolids, while it

empties

empties the ftomach and abdomen, and that it
has fomething in it that makes it preferable to
others. The new manner of giving it in fmall
dofes, does not always fucceed, and it often operates very roughly; though it muft be allowed,
that for reafons very obvious, fmall dofes have
fometimes as much effect as great : given in the
manner as I gave it, it excites the vomiting
three or four, or even eight times. This vomiting took away the ficknefs, and the more gall
there came away, the more fervice it did ; I have
even on the third day of a confirmed, though
not very violent dyfentery, with forty grains of
ipecacuanha, brought away from a farmer's
daughter fuch an aftonifhing quantity of bilious
matter, that the dyfentery was entirely fubdued
at once. With moft perfons the emetic got
away pretty much of this matter ; the excretion
of blood was commonly for a time either ftopped, or at leaft leffened, the tormina were immediately more eafy, and the ftools lefs frequent,
though this alleviation remained but a few hours.
The patients were in a very bad condition indeed, when this fhort relief did not follow at all;
otherwife this ufual confequence of the vomit
was a good prefage.

I likewife always gave an emetic with good
fuccefs, when I was not fent for till a week, a

fort-

fortnight, or more had paffed from the beginning of the difeafe, in cafe I fufpected a corrupt
matter in the ftomach, and there was neither
inflammation nor fuppuration in the bowels.
More than one I never gave; perhaps I did
wrong; but even this one was very ill taken
of me. Sometimes, compelled to it by contrary indications, I began the cure without
the emetic, and purged fo much the more violently for it with good fuccefs. To children
that were extremely young, I very wrongly gave
no emetic.

After having given the vomit in the morning, I ordered them to fet out in the afternoon
with the following drink: Take two ounces of
barley, and boil them up with an ounce of cream
of tartar, in two pints and a half of water, till
the barley burfts; then ftrain it through a linen
cloth, and fet the liquor by, which will amount
to about a quart, to be drank warm at proper
intervals, during the firft afternoon, and the
whole fucceeding night throughout. I leffened
the dofe of the cream of tartar according to the
age of the patient, though I moftly ftuck to the
proportion before-mentioned.

On the fecond day in the morning, I gave to
adults three ounces of tamarinds, boiled up for
the

the fpace of two minutes, with half a pint of warm water, and ftrained off; to children two ounces, and to very fmall infants one. This gently opening medicine directly brought on the ftools, more copioufly than before, but after this their number was generally diminifhed; fometimes the tormina went quite away, but for the moft part, were at leaft greatly alleviated. A large copious excretion produced by this me-dicine, had always an excellent effect. Inftead of tamarinds, I fometimes gave Sedliz falts to the quantity of an ounce, or an ounce and an half, with the like fuccefs. During the night, I repeated the barley-water with the cream of tartar. On the third day, I gave ftill the tama-rind decoction, if the malady was not fufficiently diminifhed; otherwife I put it off till the fourth day, and ordered nothing further in the mean time, than the barley-water with cream of tartar.

I gave the peafants pretty often after the eme-tic, on the afternoon of the firft day, a drachm of cream of tartar, with the like quantity of rhu-barb: the fame dofe on the morning and even-ing of the fecond day, and the morning of the third. Sometimes I divided this into fix dofes, and ordered the whole fix to be taken by the fourth day, while, at the fame time, I prefcribed the barley-water in the fame manner; I dimi-

nifhed

nifhed the dofes likewife in proportion to the patient's age. The fuccefs was not bad; for by means of a vomit given at the beginning, two drachms of powdered rhubarb, with the like quantity of cream of tartar, and the common barley-water, with an ounce of the fame falt; ·I have done many people great fervice in three days time, and have in this manner even perfectly cured a woman fourfcore years old, of the dyfentery. By this method, however, the pains did not fo foon remit; but on the contrary, grew much more violent; which did not happen when I omitted the rhubarb.

The cream of tartar and tamarinds did not only occafion no pain, but very much diminifhed it, when they proved fufficiently purgative. They had alfo this advantage over rhubarb, that by means of their acidity, they acted very powerfully againft the putrid fever; while on the contrary, rhubarb, except a deterfive and (as it appears to me) not very antifeptic power, can boaft of nothing more, than of being capable of contracting the fibres.

In obftinate and tedious cafes, by means of an opening medicine confifting of three ounces of tamarinds, the ftools became lefs frequent in the very height of the diforder, and the patients

were

were always relieved. So far from being weakened by this purge, I perceived that they grew stronger and more alert than they had been before, when their bowels were diftended with putrid matter.

In general, the tamarinds had a much quicker and better effect than rhubarb alone. So far from caufing pain, they alleviated it very much, and accompanied with the cream of tartar during the intervals, finifhed the difeafe in three or four days, even when the attack was very violent. Notwithftanding the emetic, the ftools grew very copious and of a bad appearance fome hours afterwards, the pains great, and the wearinefs of the members very confiderable. But very often all thefe fymptoms fuddenly vanifhed on purging the patient with tamarinds.

As faft as each fymptom of the dyfentery decreafed, and at length vanifhed, I perceived that the fever in like manner decreafed and vanifhed. It took a faft hold, and even grew very confiderable, when the putrid matter was not evacuated in fufficient quantities directly at the beginning. I made ufe of no other remedy for it, than thofe which I have already indicated. They were fufficiently capable of correcting and evacuating the bilious matter, and thus likewife of putting an end to the fever.

After

After the emetic I fometimes too gave cream of tartar, rhubarb, and tamarinds by turns, with good fuccefs. But I was guilty of an error in not being content with tamarinds, and the other medicines alone, when I had omitted the rhubarb.

A woman of fifty-fix years of age, at Brugg, went to bed in good health, and was fuddenly taken in the middle of the night with a very violent cold fit, accompanied with fome whitifh-yellow ftools and cholicky pains. She had befides a ftrong propenfity to vomit, a tafte of bitternefs in her mouth, and a perfect bilious vomit. She called me in the morning: I found her ftill in the condition I have defcribed, excepting that the cold paroxyfm was changed into a hot one, attended with fleepinefs and wanderings. I gave her half a drachm of ipecacuanha, which fhe took in the afternoon, while the cold fit was ftill abfolutely on her; this vomited her violently with much relief, and the cold fit kept off. In the evening fhe had much heat, dofing, and wanderings; as to the reft, the ftools were lefs frequent, and the cholic very tolerable. The whole night through I gave no medicine, and fhe had twenty painful ftools, which were of the colour of faffron.

The

The fecond day I gave her, at eight and eleven in the forenoon, half a drachm of rhubarb, and an ounce of cream of tartar, in a quart of barley-water, which I ordered to be drunk out the fame day. In the evening, I found the heat and fever much lefs than the preceding day; though the cold fit ftill came on at times, the ftools were pretty frequent, and painful juft before evacuation, of a faffron-yellow colour, and then for the firft time bloody. All the night long they ftill continued in great number, being likewife attended with very violent pains, and blood.

The third day, in the morning, I gave her a purge of three ounces of tamarinds. In the evening, I found that this medicine had procured eight or ten large ftools, on which enfued the greateft alleviation with regard to the pains: the fever feemed to me extremely mild. I now prefcribed for this evening and night, nothing elfe than warm panada without wine: fhe had during this night too not the leaft pain, and only one ftool.

The fourth day in the morning, I found a thick rafh on her lips, and apthæ in her mouth. I gave her for this day two half drachms of rhubarb, and repeated the ounce of cream of tartar to be boiled in the barley-water, and taken in the

the ufual manner. Thefe remedies continued to
evacuate a good deal of the peccant matter,
which was all along fomewhat bloody, but without the leaft pain : the pulfe however appeared
ftill to be a little feverifh. During the night,
the patient had three flight and fomewhat bloody
ftools ; all the reft of the time fhe refted perfectly well.

On the fifth day, I prefcribed nothing but
linfeed-tea. Her ftools entirely ceafed, as well
as her pains ; notwithftanding which fhe did not
fleep the next night, which I took for a fign of
the prefence of yet more matter, and gave her
therefore on the fixth day, three ounces more of
tamarinds in the ufual manner. Thofe tamarinds
evacuated at once an aftonifhing quantity of matter. From this time forward fhe had no more pains,
flept all night perfectly well, had a natural ftool
the following day, and was in perfect health.

I have often feen that tamarinds were of fervice, when rhubarb was not. In order to prove
this, I will at prefent relate only one cafe out of
many others.

A farmer's fon, in the diftrict of Wildenftein,
had the dyfentery in the higheft degree ; on the
fourth day I was confulted. I gave him a vomit

to be taken immediately, half an ounce of cream of tartar for the ufual barley-drink, and for the fifth, fixth, and feventh day, three drachms of rhubarb in powder, to be taken in fix times.

The eighth day his friends gave me an account, that no more blood came away from the patient, but the ftools were yet very frequent, and attended with intolerable pains : that the patient befides with each ftool, felt a violent fenfation of heat in the abdomen, and that his urine too fcalded him in a very uncommon manner, as it paffed off. I gave him three ounces of tamarinds to be taken at once, and an ounce of cream of tartar for the barley-drink.

The tenth day they brought me word, that the heat in the abdomen aud urine had fuddenly difappeared on taking the tamarinds, that the pain which remained was very tolerable, and the ftools few. I gave once more the three ounces of tamarinds, and ounce of cream of tartar, both to be taken as before. This had likewife fo good an effect, that the patient got perfectly well in a few days.

The tamarinds had in like manner a good effect, when I was obliged, for particular reafons, to leave the emetic entirely out.

A blind

· A blind man of two and fixty years, in the diftrict of Koenigsfeld, who had been long tor - mented with the arthritis, was taken with the dyfentery and putrid fever in the ufual manner; he fent for me the fecond day. I found it impracticable to give him a vomit, as he had two ruptures. I prefcribed therefore three ounces of tamarinds, diffolved in water, which he was to take directly, and an ounce of cream of tartar, for a quart of barley-water, that I made him drink out during the night.

On the morning of the third day I was told, that my patient had taken all the medicines I had prefcribed, had had an uncommonly copious and frequent evacuation during the whole night, that the pain was very much decreafed, and had grown more and more eafy at each ftool. I prefcribed again the three ounces of tamarinds, and for the night an ounce of the acid falt. On this all the pains vanifhed, my patient had the whole night through but two ftools; as to the reft he flept well, and his ftools were not longer bloody nor green.

The fifth day, I ordered my patient for the next four and twenty hours, nothing elfe than linfeed-tea fweetened with fugar-candy; that I might be able to obferve, according to my cuftom, the

difeafe,

difeafe, when left to itfelf, and queftion nature on the fuccefs of my cure. I was told on the fixth day, that during that time he had ftill had fome pain, and a few loofe ftools. I prefcribed three ounces of tincture of rhubarb prepared with water, (as he was tired of the tamarinds) and ordered him to take a fpoonful of it morning and evening. By the ufe of this, he not only got perfectly well with regard to the dyfentery, but alfo told me fome weeks afterwards, that he found himfelf better than ever with regard to his arthritic complaint, that he feemed to have loft his pains in his knotty joints, and could walk out whenever he pleafed.

The tamarinds were alfo of fervice when taken alone. A child of four years of age, in the diftrict of Caftel, had laboured under the dyfentery and putrid fever five days before I was called to it. I gave it fix ounces of tamarinds, of which it was to take two ounces diffolved in water, every morning, for three days together. By the ufe of this it was re-eftablifhed immediately, without any other help.

Acids were of great ufe. A very ftout man in Brugg, had all day long a violent fhivering, and a conftant fruitlefs endeavour to vomit. In the evening he was feized with a violent belly-
ach,

ach, which lafted almoft without interruption all
the night following, and was accompanied with
frequent and copious ftools. The fecond day
he called me in.   I gave him half a drachm of
ipecacuanha, which vomited him twice with
great relief; the pain though very violent, returned but feldom throughout the whole day,
and he had about twenty ftools.   In the evening, I gave him half an ounce of cream of tartar, to be taken directly, and continued the whole
night throughout, with a quart of barley-
water.   He drank this out, and his pain and
ftools difappeared entirely by the morning. The
third day I gave him three ounces of tamarinds,
that brought on three gentle ftools: on the
fourth, he was well.   But in a much more confiderable cafe of the dyfentery, I have perfectly
cured a perfon fixty-eight years old in four days,
by the ufe of three ounces of tamarinds, and half
an ounce of cream of tartar every day.

I could not however always bring the affair to
pafs, purely with evacuating and antifeptic remedies.   The pains were at times extremely
violent, when the patients had not been purged
enough at firft, and in the courfe of the difeafe
were too averfe to taking the opening medicines;
accordingly the bearing down was in thefe cafes
intolerable.   I was alfo fometimes obliged to
think

think of medicines more immediately anodyne, and where my hands were bound in various ways, to leſſen the too great violence of the purging by innocent means.

I always looked upon it as dangerous to give opium in the dyſentery, before the fuel, that fed the fire of the diſorder, was burnt out. I endeavoured therefore to find out a method of giving opium, in theſe obſtinate and extraordinary painful caſes, with as little prejudice as poſſible. This happened for the moſt part to be of benefit, but was not always without prejudice.

The laudanum liquidum Sydenhami, given to ſix drops in linſeed-tea every ſix hours, to a pale young gentleman about nine years old, who for ſome years had been plagued with worms, ſoothed, indeed, his violent pains on the eighth day of the diſerder, after a hearty evacuation, but it very much increaſed his fever; though at the ſame time I ordered him every three hours day and night, a large ſpoonful of tincture of rhubarb prepared with water. It cauſed likewiſe, in this child, an endeavour to vomit, as it retained the putrid matter ſtill remaining behind, and indeed brought on a downright vomiting. But all theſe indiſpoſitions vaniſhed on the repeated uſe of the tama-

rinds,

rinds, cream of tartar, powder of rhubarb, and on totally laying afide the laudanum.

Sixteen drops of the laudanum Sydenhami, given to a young man of Brugg, after copious evacuations made on occafion of his being afflicted with violent pains in the bowels, excited anxious dreams, and a pretty fmart pain in the joints, while that in the belly ceafed entirely; however, this pain in the limbs vanifhed the next day. Eight drops in the evening, and the fame quantity at midnight, were on the contrary afterwards of good effect in the very fame perfon; he had no pains in the joints, nor in the belly, no dreams, lefs fleep, and feven ftools during the whole night, inftead of an hundred and fifty or two hundred, which he had had before every night. Yet the diftemper grew long and tedious, and continued on him a fortnight; this I afcribed to the laudanum, which to be fure eafed the patient, but by the very eafe it gave, lengthened the diforder. This is the only one among all my patients, who had the prolapfus recti at the end of the difeafe; but he foon recovered, and remained from that time perfectly frefh and healthy.

In four other cafes I have obferved, that the laudanum Sydenhami given after proper evacuations,

ations, alleviated the pains, and leffened the number of ftools, without ftopping them entirely; I
began then immediately with the rhubarb.
Whence appears, that it fometimes had the advantage of fomewhat diminifhing the ftools, without putting a ftop to them, and in the mean time
of taking away the pains : but without rhubarb
given in the intervals, or juft after, it was very
plainly pernicious.

In the child of a perfon of condition, a year old,
notwithftanding all its crying and purging, the
dyfentery was not difcovered till the fourth day,
and then only from the circumftance of the pure
blood being feen run in ftreams down its thighs.
After the moft violent pains, this child fell into a
continued lethargy, with perpetual fpafmodic
contractions in all parts of the body : I expected
nothing elfe to follow, but death. In the mean
while I gave him a purge of tamarinds, every
three hours the whole day and night throughout, two large tea-fpoonfuls of the tincture of
rhubarb made with water, a great deal of linfeed-
tea, and every fix hours, three drops of Sydenham's laudanum. Its excrements were very copious, and of various colours, white, yellow, brown,
green, red, and black. By this method the child,
in fpite of a miliary eruption, (which appeared

F   · towards

towards the end of the diforder, and went off fponta-
neoufly by defquamation) got well in a fortnight.

To a child of two years of age, in Brugg, that
upon the firft onfet of the diforder, entirely loft
all fenfation from a fpafmodic contraction of the
nerves, I gave the tamarinds and the tincture of
rhubarb, but no laudanum; accordingly it died.
The death of this child is the only one, which I
could afcribe to my want of capacity and knowledge in my profeffion; the few other deaths
that happened, were entirely owing to the un-
towardnefs of my patients.

I found chamomile-tea the beft means next
to opium, of alleviating the pains, and befides
that, it is at the fame time antifeptic. I gave
this infufion, which is even of fervice in inflammations of the bowels, in great plenty, and very
often with good fuccefs. Linfeed-tea, rice-
gruel, barley-water, and clyfters of gum arabic,
were alfo very ferviceable in mitigating extraordinary violent tormina; though the clyfters
came away often without effect, and therefore I
could place but very little confidence in them at
the height of the diforder. I likewife gave an
almond-emulfion to be drunk quite warm, which
the patient found very ferviceable, whenever the
pains were confiderable.

In

In general I perceived very well, that the pain
could not be radically cured by any means, unlefs
the fole caufe of the diftemper, the putrid matter, were difcharged. In like manner the
violent tenefmus, which was fo very troublefome even at the end of the diforder, was not
put a ftop to either by Diafcordium clyfters, or
the Theriaca and milk, as Huxham advifes; for
the evacuations muft be repeated, as long as the
tenefmus continues. In many of my patients,
I found at the end of a very violent illnefs, a
painful irritation to ftool, which for the moft
part was fruitlefs, and which was followed very
feldom by a very fmall ftool. I accounted for
this, from the lofs of the natural mucus in the
rectum. But this explication was falfe; for I
gave them in confequence of it, clyfters of half
an ounce of gum arabic diffolved in warm water,
and they were of no fervice; I gave laudanum,
and it was juft the fame; I then gave them
morning and evening, a fpoonful of the tincture
of rhubarb above-mentioned; and this was of
the greateft benefit. Whence I learned, that
this tenefmus, which appeared towards the end
of the difeafe, did not proceed from the too great
denudation of the inteftines, and an increafe of
their fenfibility proceeding therefrom; but in
reality from a fubftance, that ftill remained in
the cavities of the great guts.

Many

Many of my patients called me in late, and sometimes extremely so. In these cases, where Simaruba, Cascarilla, and the Terra Iaponica are usually thought to be absolutely necessary, I gave sometimes an emetic, and with rhubarb alone finished the whole affair, when the disorder had lasted very long. A woman of sixty-three years of age, in the district of Wildenstein, that had had the dysentery a week, and had at that time fifty stools in twelve hours, at the same time that every thing she eat and drank came away from her by vomiting, I cured with an emetic, and the other remedies above-mentioned, in a few days. I have made some people well in the country, that had been a month together without medicines, and were attacked with a long tedious dysentery, attended with a great sense of weariness in the members, shiverings, violent sweats, and a perfect indigestion. All that I gave them, consisted merely in a few doses of rhubarb in powder, to the quantity of half a drachm, which I ordered them to take in the morning for two days together, with chamomile-tea. With this powder they were tolerably well purged and with great benefit, getting well in a couple of days; on the contrary, these cases were dreadfully prolonged, when in contempt of physic and physicians, the disorder was left to nature, quacks, and old women. A woman in the

district

diſtrict of Caſtel, had had the dyſentery already
ten days in a very high degree, when ſhe con-
ſulted me. I gave her only for a couple of days,
ſome cream of tartar to be diſſolved in barley-
water, the rhubarb powder, and chamomile
flowers for tea, in expectation that ſhe would ſend
me an account of the manner in which ſhe found
herſelf, and aſk for more medicines. She ſet
thoſe ſhe had aſide, as they were not of ſervice the
very firſt day, and let me hear no more of her.
She continued thus with the dyſentery on her for
the ſpace of nineteen weeks, following a thouſand
old womens recepts, and at the end of that time
I learned by chance from her huſband, that ſhe
had not yet got rid of her complaint, and ſtill
had blood in every ſtool.

To ſuch, as without keeping their beds, had
only a violent belly-ach, and at the ſame time
were coſtive, I gave for ſome days together a
drachm of powder of rhubarb, to take at twice.
The excrements were likewiſe in thoſe perſons,
immediately after the firſt evacuation, bloody,
and white, as if they were purulent. But after
ſome ſtools they found relief, the tormina vaniſhed, and they were well in a few days.

Almoſt all the ſick, that were cured by me to
a conſiderable number, ſhewed an extraordinary

appetite

appetite while under cure.   For which reafon, I
did not think any medicines to ftrengthen the
ftomach and bowels here neceffary.   Eating
ftrengthened them enough : to fome I gave, by
way of corroborating the ftomach, a fpoonful of
tincture of rhubarb every morning; to others,
Hoffman's vifceral elixir.   But I never gave
corroborating medicines merely with a view of
hindering a relapfe; for I do not think, that
even our egregious Mr. Conrad Rahn *, can
prevent the dyfentery by means of ftomachics.

In general, at the beginning of the diftemper,
ipecacuanha, cream of tartar given in great quantities with barley-water, and tamarinds, were my
principal remedies.   Againft the tormina, I
made ufe of chamomile and linfeed-tea, almond-
emulfions, gum arabic clyfters, and, but feldom,
and that with the greateft caution, of laudanum.
Towards the end of the difeafe, rhubarb was of
fingular fervice.

---

* The reader will be pleafed to obferve, that what is
faid here and in the following pages, of this gentleman,
is by no means to be taken in a ferious light.

CHAP. V.

The Effects of other REMEDIES.

WE muſt try many things, obſerve and compare every thing; if we would not be mere novices in the knowledge of nature, and would wiſh ſometimes to draw conſequences from our obſervations, that may be of univerſal uſe, and extenſive in their effects.

We grow ſometimes rather negligent from ſome ſuccefsful events, and do not examine, when the ſick are cured, whether we could not have ſaved them by a ſhorter, better, ſurer, and more univerſally uſeful method. This ſelf-applauſe in a phyſician, is a moſt invincible obſtacle to his progreſs in the profeſſion ; for when one meets with approbation, one ſhould always conſider, whence it comes. I honeſtly confeſs, I did not treat the firſt patients I had in this year's dyſentery, with as great depth and ſolidity of judgment, as thoſe that came after ; though not one of my firſt patients died : notwithſtanding which my method was faulty.

In the fifteen years that I have practiſed phyſic, I never ſaw the dyſentery ſo frequent as this

year.

year. Notwithſtanding which, I have during
this long ſpace of time, viſited a tolerable number of dyſenteries, and followed a method with
them, that was ſo far not unfortunate, as none
of my patients died; and of which I will give
the two following examples.

A woman of ſixty-one years of age, worn out
with frequent hyſteric fits, and who had been
before this very often attacked with many violent diſtempers, was ſeized in the year 1759,
with a terrible dyſentery ; ſhe ordered me directly to be called.   I preſcribed her the tincture of rhubarb prepared with water, a large
ſpoonful to be taken every three hours day and
night, and in the intervals I made her drink
plentifully of an almond-emulſion, in which gum
arabic was diſſolved : I likewiſe ordered them
often to give her clyſters of gum arabic and
barley-water.   By degrees her intolerable griping pains, her very high fever, and her teneſmus diſappeared, on the uſe of theſe remedies :
her ſtools too decreaſed very much.   At the
approach of the fourth night, I thought I might
venture to give ſixteen drops of the laudanum
Sydenhami : this night ſhe was very quiet.   The
fifth day the patient was eaſy, without pain,
ſtools, bearing down, or fever, and quite chearful and hearty.   I now gave her no more medicine,

dicine, in order to fee, whether the laudanum
did not deceive me. But likewife in the afternoon all the bad fymptoms kept off, excepting
that fhe was a little delirious. I gave her neverthelefs the almond emulfion with the gum
arabic in the evening, and things remained as
they were before; in a few days, by means of
fome ftrengthening medicines, fhe was perfectly recovered. Two years after fhe had the dyfentery again, and that very violently; in a week I
fet her up again, with the very fame remedies.

A very ftout, healthy, frefh, lively lad, about
twenty years old, brought with him in the year
1763, the dyfentery from Zurzach, where it
then was epidemic. It was very violent, and
attended with all kinds of bad fymptoms: the
fick perfon's father, a country parfon, and a fage
of the Paracelfian fchool, gave him infallible remedies againft the dyfentery one after another,
while in the mean time the diftemper rofe higher
and higher. I, a fimple layman, was called on
the eighth day of the difeafe, and found the lad
quite fpent and overcome, with a down-caft,
fhrivelled, and perfectly corpfe-like countenance,
though before he was very handfome in the
face. His voice was flow, weak, and broken;
the cold fweat ran from him on all fides, and
bloody gangrenous ftools came away with violent
pains

pains every minute.   I told the unhappy father
to throw all his infallible remedies out of the
window; and gave the fon, on the eighth,
ninth, and tenth day of his diforder, which
was now in reality rifen to a very high degree,
not fiery cordials, and infallible remedies that
would have killed him, but only ftrong dofes
of the tincture of rhubarb in water, almond-
emulfions, (which at firft he brought up again)
barley-water, and gum arabic clyfters.   By this   •
moft unchemical, and in the higheft degree,
fimple method, the youth that before lay at
death's door, obtained in a few days his former
health, and foon after his full ftrength, his ufual
chearfulnefs, and his fine lively colour.

Thefe examples, felected from many others,
induced me to try alfo the fame method, or one
very little different from it, in this year's dyfentery.   I will at prefent cite two examples of its
fuccefs.

A woman at Brugg, thirty-feven years of age,
of a very irritable temperament, and fubject to
the moft violent hypochondriacal and hyfteric
fymptoms, was feized this year in June with the
dyfentery. She fent for me on the third day, and
and told me, that fhe had hitherto had only twenty
ftools in twenty-four hours, but that thefe were
accom-

accompanied with very violent pains, and that with each ftool fhe had a propenfity to vomit. I gave her three drachms of rhubarb in powder, divided into fix parts, and ordered her to take one of them every two hours, with chamomile-tea, between whiles water juft boiled up with rice, and almond-emulfion. The fourth day I found my patient had lefs pain, her ftools were as copious as before, and ftill tinged with blood. I gave her again fix half drachms of rhubarb, to be taken in the fame manner as the firft. In the evening fhe was fomething better, fhe had had in the mean while feven ftools more: I ordered her at night nothing but the rice-water. The fifth day the fick woman 'told me, fhe had had in the night eight very painful ftools, and an intolerable tenefmus. I prefcribed her a fpoonful of tincture of rhubarb every three hours, half an ounce of gum arabic diffolved in half a pint of water, to be taken by way of a clyfter directly, and a fecond clyfter of the fame kind in the evening. Upon thefe medicines her pain almoft entirely ceafed, and fhe had the whole day long but two ftools, and thofe without blood. During the night I made her drink almond-emulfion, and the rice-decoction. The fixth day fhe found herfelf very well; however, I let her ftill go on with the tincture of rhubarb, and the fame diet. She recovered her health,

but

but neverthelefs had two relapfes; the firft from being put into a paffion by a quarrel, the fecond from being twice wetted through at night with the rain. She was cured by the fame means as before.

A fickly, heavy, lethargic perfon, of nine and twenty years, inclining to green ficknefs and white fwellings, was attacked very violently at Brugg this year in Auguft, with the dyfentery. Being applied to for advice the fame evening, I gave her an ounce of tincture of rhubarb prepared with water. She had the whole night throughout very violent pains, a great many ftools, and a conftant propenfity to vomit. I gave her the fecond day half a drachm of ipecacuanha, and let her afterwards go on with the tincture of rhubarb as before; as to the reft, I allowed her to eat or drink nothing but ricegruel, water-gruel, and chamomile-tea. In the evening fhe told me, that a vaft quantity of mucus and bile was come away with the vomit, that her tormina were violent, and her ftools not very copious: as to fever, I did not find that fhe had any, and gave her an ounce of tincture of rhubarb to be taken as before. In the night her ftools were more frequent, and her pains almoft intolerable. She was in the fame ftate on the morning of the fifth day, when I
ordered

ordered her an ounce of the tincture of rhubarb,
and two half ounces of gum arabic for as many
clyfters. But thefe clyfters came away immediately each time, my patient was all day long
obliged to go to ftool every minute, with in-
effable torments, and there came away an im-
meafurable quantity of water, mucus, bile, and
blood. I bid her take an ounce of rhubarb as
before, and drink almond-emulfion plentifully,
with rice-gruel. Notwithftanding which, fhe
had juft in the fame manner a large ftool every
minute, with cholicky pains. Her nurfe could
not bear any longer the fmell of her ftools,
though fhe aired the chamber continually, and
carried them out immediately, as faft as they
came away.

The fourth day in the morning, all thefe
complaints were at a very great height. I prefcribed, together with the ufual drink of almond-
emulfion and rice-gruel, two fpoonfuls of a
mixture compofed of an ounce of tincture of
rhubarb, half an ounce of gum arabic, and feven ounces of water, to be taken every hour.
In the evening I found my patient fitting almoft continually on the clofe-ftool, her excrements inexpreffibly copious, enough to ftink
one to death, yellow, green, brown, black, and
ftreaked with a quantity of blood; fhe at the
                                                    fame

fame time full of pain, anxiety, and terrors almoft to defperation, extremely weak, and at times
even fainting away. I let her take all at once
half of the mixture ordered in the morning,
and which fhe had not till then made ufe of.
I then gave her for ten o'clock at night, twenty
drops of the laudanum Sydenhami, and ordered
that fhe fhould drink as much almond-emulfion
as poffible all night long. Her dead pale
cheeks were affected with a great heat, as foon
as fhe had taken the laudanum ; fhe flept fome
hours afterwards, had a few ftools, and very
little pain. On the fifth day in the morning, I
made her take the other half of the mixture,
with the almond-emulfion and chamomile-tea.
In the afternoon I ftill found her with the flufh-
ing in her face, a fly undermining fever, and
but few ftools in the fpace of an hour, which
were no longer of that cadaverous quality as before, nor bloody, but attended with great pains.
I ordered that a clyfter fhould be given her of
gum arabic and water, which came away directly without effect. In the evening her pains
were very violent, at night I gave her fixteen
drops of Sydenham's laudanum, and prefcribed
a good quantity of almond-emulfion. She flept
fome hours, had five ftools in the night without
pain, and in the morning found herfelf better
in all refpects.

                                                The

The fixth day, I ordered nothing elfe in the morning than the fame clyfter, rice-decoction, and almond-emulfion ; fhe had three ftools, and her excrements were again fomewhat red. In the afternoon I found my patient without fever, heat and pain : I prefcribed again a clyfter of gum arabic, and the fame drink. ' Notwith-ftanding which the pain returned, her ftools con-tinued, and were green, black, and fomewhat tinged with blood, but much lefs ftinking than before ; fhe herfelf was quite free from any fe-ver, but feeble, and inflated : I gave her fixteen drops of laudanum. This medicine had the ufual effect in the night, when fhe had three ftools in the fame manner as before. The feventh day I gave her early in the morning an ounce of tincture of rhubarb, of which fhe was to take a fpoonful every two hours, and as to the reft, to go on with the fame diet-drink as before. The confequence was, that the patient had that day nine perfectly yellow, ineffably ftink-ing, but in no wife painful ftools, which gave her the greateft eafe. At night I gave her fix-teen drops of laudanum ; fhe had two ftools, almoft without fmell. The eighth day I gave no medicines, to fee how far the diforder had really gone : fhe had a few, bilious, painful ftools, that began already to fmell ftrong. I prefcribed for the night, laudanum, and for the

following

following day, every three hours a spoonful of tincture of rhubarb. The ninth day she found herself perfectly easy, was hearty, and contrary to her usual custom, in good humour : her stools were very few in number, but still continued bilious ; I made her go on with the tincture of rhubarb, and advised her now to eat something more solid. In the night she had some pain, and on the tenth day five stools, otherwise she was very well. It was the same on the eleventh, when I prescribed her only a spoonful of the tincture of rhubarb morning and evening ; she slept the whole following night, and her loosenes had entirely ceased. Things were in the same state on the twelfth day, notwithstanding which I advised her to take the tincture of rhubarb twice a day. On the fifteenth day she was so well recovered, as to complain of nothing but some remaining weaknes ; I gave her an ounce of the elixir of vitriol, to take forty drops in water twice a day, with which she got perfectly well.

A person that has any knowledge in physic, will easily perceive the faultines of this method, with respect to the dysentery of 1765. Many physicians think themselves very bold, when they venture to give the tincture of rhubarb in a few drops to adults ; which, to me, has the look of a man's attempting
ing

ing to make a breach in a battery, with the fame
fhot as he ufes to kill fparrows with. Degner
gave in the dyfentery at Nimeguen but half a
fpoonful, or a fpoonful, every four or fix hours.
My dofes were larger and more numerous; notwithftanding which, the effects of the rhubarb
were too flow and tedious, as it was neither purgative nor antifeptic enough, and let the diforder get to a great height. In fome very fe·
vere cafes of the dyfentery, which on account of
the prolixity I have been guilty of in other
places, I have not here related, I gave at firft
an emetic, and afterwards till the fifth day tincture of rhubarb in great quantities, with al-
mond-emulfions, and rice-water, without any
good effect: on the other hand, cream of tartar
produced a fudden alteration for the better, by
rendering the ftools more copious, and of a more
folid confiftence. Hence I concluded, that this
method was in general of no fervice in this epidemy; that I might for the moft part referve the
tincture of rhubarb for flight indifpofitions, (in
which indeed I commonly found it of benefit)
but that it was an excellent remedy towards the
end of the diforder.

In like manner, the rhubarb in powder did
not purge fufficiently in the beginning. It
always increafed the pain, (which however did

G   not

not happen on taking it in tincture,) and the number of ſtools was not very much leſſened. Mixed with cream of tartar it purged better, though ſtill with much pain ; on the contrary, the tamarinds evacuated quickly, copiouſly, and without bringing on freſh pain ; and directly after their operation, the number of ſtools was ſuddenly diminiſhed.   Thoſe perſons to whom I gave an emetic in the morning, and afterwards that evening together with the ſubſequent morning and evening half a drachm of rhubarb, always recovered ſlower than thoſe, to whom I preſcribed at the ſame time a great quantity of cream of tartar with barley-water.   We may ſee from all this, that though many great phyſicians, and even Degner himſelf, look upon rhubarb as the choiceſt purge in nature in dyſenteric caſes, on account of its poſſeſſing at the ſame time a laxative and corroborating, or rather an aſtringent power, they did ſo without ſufficient grounds : that rhubarb in a dyſentery attended with a putrid fever, without the addition of acids, lets the diſorder go on its own train ; and thus is the occaſion of its being prolonged, and is therefore by no means a ſpecific in this diſtemper.

Being of a rather incredulous diſpoſition, and acquainted with one particular only, a method that anſwered my views in every reſpect, I was not inclined

clined to try a heap of other remedies, otherwife very much prized in the dyfentery. On the other hand, two other phyficians have communicated to me fome very important experiments on the ufe of the vitrum ceratum antimonii, grapes, and other fruit, which I will here relate with the moft unfeigned pleafure, for their honour, and the benefit of my fellow-creatures.

A Lutheran clergyman, of great penetration and underftanding, the Rev. Mr. John Merk, minifter at Ravenfburg in Swabia, encouraged Dr. Moehrlin, a phyfician in this city of great ingenuity, and twenty-eight years practice, to make a trial of the vitrum ceratum antimonii. Eight days afterwards this phyfician told him with a fmile, the meaning of which a man, who has any fentiments of humanity, cannot be ignorant of, that he had tried the medicine on three perfons ; but had occafioned therewith fo great indifpofitions, that he was in the greateft hafte to have recourfe to every thing he could think of, capable of putting a ftop to them, and had therefore no great inclination to try it any further. But as Mr. Merk faw very well, that this remedy could not have ill performed the intended effect, in regard to the diforder itfelf ; he begged the phyfician with the irrefiftible argument of the benevolence due to his brethren, not yet to give up the

affair,

affair, the rather, as the concomitant effects
could be eafily remedied. Some weeks after,
the doctor faw Mr. Merk again, and told him
with the greateft joy, that he had already faved
fome perfons by means of the vitrum ceratum,
the acrimony of which was obtunded with
marfh-mallows root, and in the fpace of two
days had very well recovered fome, on whom
the ufual remedies had proved ineffectual, who
had been even delirious, and had one foot in
the grave. He therefore continued this remedy,
efpecially where any thing malignant appeared,
and that always with very good effect.

Upon this, Dr. Moehrlin was pleafed to fend
me a circumftantial account of thefe cures from
Swabia. He made the firft of the above-mentioned experiments, upon a woman of feventy
years of age. He gave her fix grains in lukewarm water in the morning fafting, and ordered
her not to eat or drink any thing for the fpace
of three hours ; after that he went again to fee
her, but found her quite weak and miferable,
and apprehended a fatal and fudden end to the
diforder : however he encouraged her, and gave
her with his own hands a good portion of fat
mutton broth. In the fpace of two hours fhe
had twenty ftools, after that the excrements
were without blood, the cholicky pains went
off,

off, and she rested well two hours in the night.
The following day her loofenefs was still more
diminished, and Dr. Moehrlin in the Hippocratic method, stood still without doing any thing.
The third day the woman thanked her physician
for his excellent remedy, and told him, that she
had slept well the whole night through, and had
scarce had three stools. Upon this the doctor
ordered her nothing else than a good diet, and
after some days found his patient perfectly set
up again.

Hereupon Dr. Moehrlin continued the use of
this remedy, and the rather, as at the end of
August the number of the sick daily increafed,
and rhubarb and simaruba were too dear for the
lower sort of people. The first dose of six grains
occasioned sickness and faintings with every body;
the second and third however did not; yet pre-
judiced people were much against the antimony,
as the apothecary made them believe ill of it, because in truth he gets very little by it. However Dr. Moehrlin took it into serious confideration, whether he could not remedy these bad
effects: this was not very difficult, for instead of
making use of his Æfculapian authority to forbid all sort of drink, he had nothing else to do,
than to order directly at the beginning of the
disorder, a bason full of barley-gruel, or the like

emollient

emollient drink.   At length he took it into confideration, whether it would not fucceed to mix
three or four grains of marfh-mallows root powdered, with fix grains of the antimony ;   nature
now effected what the doctor defired,   the ficknefs and faintings did not come on, the excretions were quicker, more copious, and without
pain.

    Dr. Moehrlin had given this remedy to above
feventy perfons of all ages :   there were but few
that required more than three dofes, (each dofe
containing fix or eight grains) in order to be
cured: the firft dofe increafed the flux, with the
fecond it diminifhed, and with the third it difappeared.   It was very feldom neceffary to increafe the quantity of the above-mentioned dofes,
or to add to the number of them.   With one perfon, who did not follow the diet prefcribed, Dr.
Moehrlin proceeded as far as the ninth dofe,
the laft of which confifted of fourteen grains;
this in the fpace of four hours occafioned above
thirty ftools, and upon that the tormina, tenefmus, and even the fingultus ceafed entirely, fleep
returned, and after fome days the patient was
perfectly well.   Dr. Moehrlin found that blood-
letting was one of the beft preparatives towards
the cure of this diforder, when he was called directly at the beginning; on the contrary, he
found

found it of great diſſervice, when the malady had
already ſpread itſelf through the whole body.
In this caſe three doſes were not ſufficient, and
even then at the end of the diſorder a dropſical
tumor extended itſelf over the whole body,
which continued for many weeks.

All theſe things being taken together, we muſt
confeſs that this famous antimonial remedy, long
ſince made known to us by the Edinburgh eſ-
ſays, ſhewed itſelf in Ravenſburg, to be one of
the beſt and choiceſt remedies againſt the then
epidemic dyſentery.   I ſhall have occaſion in the
tenth chapter, to take it more fully into conſideration.

A much more agreeable remedy, but which
in conſequence of a generally received, though
very ſilly notion, was looked upon as a real poi-
ſon, was made trial of with the ſame good ſucceſs.   Dr. Keller, an ingenious, careful, and extremely diſcreet young phyſician, at Weinfeld in
Torgaw, had not only had an opportunity to
vindicate the innocence of grapes and fruit, with
reſpect to our dyſentery, but even to experience
in the cleareſt manner, their prodigious uſe in
the cure of this dyſentery.

He began with a child a year and a half old, which had been haraffed with the dyfentery for the fpace of eight days, in a miferable manner. Whatfoever artifice was employed, it could not be brought to take the leaft medicine, had already had feveral convulfions, and appeared to be very near its end. The parents in the greateft diftrefs, begged of Dr. Keller in the moft urgent manner, to try every thing in the world to fave their child. He advifed them to give it grapes: the parents fears for its death, got the better of their prejudices, which might otherwife have occafioned it; the child eat two grapes in the evening, and flept very found the whole night. The next day they gave it more grapes, and after having eaten a great many of them during the fpace of eight days, it got perfectly well.

In the mean time Dr. Keller had under his care a gentleman of rank, and of great knowledge in medicine, who was attacked with the dyfentery: this perfon had taken the neceffary evacuating medicines with good effect, but could not refolve upon taking any other remedy. In thefe fad circumftances Dr. Keller advifed him in the fame manner the ufe of fruit, and three days afterwards received from him the following letter.

SIR,

SIR,

" AN averfion for all forts of medicine has
" brought me at laft to a refolution of try-
" ing the effects of fruit. In the forenoon I be-
" gan with two grapes; at noon I indulged
" myfelf with fome frefh plumbs boiled; after
" that I eat fome more raw, at the fame time
" three peaches, and towards evening a few
" black-berries. All this agreed very well till
" eight o'clock, when the tumult in my bowels
" began, and continued on in the fame way till
" midnight, fo that in all that time I could fcarce
" ftay half an hour in bed; though all this was
" without pain, tenefmus, or any other indif-
" pofition. Without doubt two potions of
" manna, and four rhubarb powders, would not
" have had fuch an effect upon any man in the
" world. On this followed my natural reft; in
" the morning I found myfelf very well, and
" eat my hafty-pudding with pleafure. This
" happy fuccefs encouraged me to go on in the
" fame manner the next day. The effect was
" exactly the fame, though fomewhat lefs fud-
" den. My appetite and fleep increafe, and
" every thing goes on better, thank God, from
" day to day."

This

This letter, which Dr. Keller inceffantly read to all he met with fick or well, prevailed at laft fo much, that every body had a mind to eat fruit; and that every one, that was not obftinate to the higheft degree of folly, did in reality eat it, and all with the beft fuccefs.

A phyfician of penetration, but by reafon of the milkinefs of his difpofition fomewhat timorous, and perhaps not free from all kinds of prejudice, has fince faid in the prefence of the fociety for natural enquiries in Zurich, That fruit, though he could not but allow it to be of real ufe in the dyfentery, might neverthelefs do harm, on account of the quantity of air which it generates, according to Hales's and Macbride's experiments, by diftending too much the fibres of the inteftines, already violently irritated by the diforder. Hereupon Mr. Chamberlain Heidegger of Zurich, a great ftatefman, who with the moft inconceivable penetration pierces into every thing worthy of being known under the fun, replied, that this was really true; but the fame experiments fhewed, that the air generated was again abforbed on the ceffation of fermentation, by the juices of the fruit, and therefore this effect could not laft long. It appears to me, that grapes and other fruit inflate bowels that are very weak, and otherwife fubject to too great expanfion, when they do not prove

purga-

purgative, and the body is rather bound, or the tenefmus very great; juft as manna is very inflating, when it does not purge fufficiently; but by the purgative power of certain fruits, and principally grapes, the ftools of moft of my patients were promoted, and with them the wind moftly came away. Confequently it was juft as little neceffary to fear in this cafe, that the fick would burft afunder like a bomb, from eating a few grapes, as it was for the great Winflow to proftrate himfelf before the altar of the virgin Mary (when he had prefcribed two ounces of manna in the hofpital at Paris) under an apprehenfion that it might excite too ftrong a purging, or tear in pieces the tender tiffue of the fibres of the inteftines.

It follows from the contents of this chapter, that in reality the tincture of rhubarb prepared with water fometimes performs very good cures in the dyfentery; but that this tincture in general, was too feeble a remedy in our dyfentery; that the powder of rhubarb, without the addition of other remedies, let it go on its own train, and thus was the means of prolonging it, and that therefore rhubarb is no fpecific in the dyfentery; that the vitrum ceratum antimonii, fhewed itfelf to be one of the beft and choiceft remedies in this diftemper; and that grapes and

fruit

fruit likewife evidently appeared to be an excellent remedy for the fame, though in the Hippocratic affemblies of old women, it ftill meets with a very bad reception.

C H A P. VI.

Effects of Aftringent and Conftipating Medicines, of Aromatics, Brandy, and Wine.

THE old phyficians fo far agreed with one another in the cure of the dyfentery, that inftead of ftriving to evacuate the matter, they chofe unanimoufly rather to reftrain it, and put a ftop to its efflux, with incraffating and aftringent remedies ; by thefe indications the diet and the whole method of cure was regulated.

Such notions, as are the produce of ignorance and folly, are never to be eradicated. It is true, the phyficians of our times, oppofe for the moft part that method, which is calculated to cure the bilious dyfentery, by the means of conftipating and aftringent medicines ; but mankind are too apt to reject in fpeculation, what they make ufe of in practice. Aftringent remedies are as yet by no means banifhed, and are prefcribed by ninety phyficians out of an hundred ; it is true, they begin with a few evacu-

ating

ating remedies; but of what ufe is it, to give the
firft day an emetic, the fecond rhubarb, and afterwards nothing elfe but conftipating and aftringent medicines. I found myfelf forced twice in
our dyfentery of 1765, to prefcribe a purge even
at the ninth and eleventh day of the diforder;
as my patients were then in the greateft danger,
their fever very violent, their ftools innumerable,
and their weaknefs uncommon. This purge confifted of tamarinds; after the evacuation, which
followed it, the ftools were very much diminifhed
in quantity, the patients heartier than before, and
the diforder was at an end in a few days. Suppofe
now that I had done in thefe cafes, what moft
phyficians generally do, that is, after the evacuations of the firft day, gone on with reftringents:
without doubt the confequence would have been
a tedious fit of ficknefs, or death.

My hair ftood an end, when the Provifion (as
it was called) againft the dyfentery, made by
the excellent college of phyficians of Bern in
1727, and diftributed among the country people, lately fell into my hands. At the very beginning of it, the phyficians of Bern make this
remarkable obfervation, that the then epidemical dyfentery did not perhaps only proceed from
a difordered ftomach, but fhould be confidered
in the light of an acute fever, which produced

an

an inflammation of the bowels. We may hence
conclude, that that species of the dysentery, which
is accompanied with an inflammatory fever, was
at that time epidemic there; and yet the physicians of Bern in their Provision, prescribed almost
nothing else than extremely constipating and
astringent medicines; consequently every thing
in nature, that had the greatest power of increafing the inflammation.

The Provision (as it is called) against the dysentery, issued out by the very same college in
the year 1750, which was printed and distributed
through the whole land, is, it must be owned,
something changed; however, excepting a very
large dose of ipecacuanha, and a couple of doses
of rhubarb, the other medicines are as constipating and astringent as possible. Without doubt
this method was at that time good in many
cases : but I hope to be excused for not giving
credit to this Provision, in the epidemy of
1765, although it was dealt out by I know
not what mistake this year, in several places
up and down the country. Some practitioners gave credit to it, and followed it so well,
that their patients could certainly not have recovered in less than a quarter of a year. But
I fear on this account, that the light of this
century does not shine upon these gentlemen,
who

who have perhaps never read any other medical work in their lives, than this fame Provifion.

The fpirit of contradiction is not my failing;
as much a martyr to the truth as I otherwife am
in this my beloved country. In the mean while
it is and ever will be certain, and very eafy to
demonftrate, that aftringents (or opiates given
before the proper time) in the dyfentery, retain
the venomous matter in the body; increafe the
pain, fever, heat, and danger; excite anxiety
about the precordia, the hiccough, ulcers in the
mouth, vomiting of blood, inflammations in
the bowels, and a mortal gangrene; or throw
the patient into a continual cholic, with an obftinate conftipation of the bowels, into the gout,
hectic fever, jaundice, tympany, œdematous tumors, the dropfy itfelf, and even abfolute lamenefs.
However, I rather chufe to leave it to Dr. Degner and Tiffot, to exprefs my fentiments on this
fo much admired aftringent and conftipating
method of cure.

Degner fays; " In the dyfentery at Nime-
" guen, the empirics were unanimous in mak-
" ing it their fole endeavour to put a ftop to,
" and retain the alvine flux, little caring whether
" the morbific matter were fufficiently evacuated
" or

" or corrected. The too early and copious use
" likewife of narcotics and opium, had a bad
" and even mortal event, as the remiffion of the
" pains and flux arifing from thence was very
" deceitful. The phyfician and his patients
" grew too fecure, and the hoftile attacks of the
" malady were lefs vigoroufly withftood, than
" they ought to have been ; whilft one funk into
" an eternal fleep, another died of an inflamma-
" tion of the bowels, and with others the loofe-
" nefs ceafing for a few hours or days, returned
" afterwards more violent than before." In another part of this excellent phyfician's work, I
find he is of opinion, that thofe who give aftringents in the dyfentery, keep the fnake fhut up in
the patient's bofom, by hindering the efflux of
the acrid peccant matter out of the body,
whence proceed inflammations, ulcers, gangrenes,
and death.

Tiffot fays; " The very worft method of
" cure, is juft that which is moft followed. The
" evacuations are cohibited either by aftringents
" or opiates ; a fatal method, that fweeps away
" yearly a great number of perfons, and throws
" others into incurable diftempers. By retaining
" the excrements, the wolf is fhut up in the fheeps-
" fold. The matter that remains behind, irri-
" tates and inflames the inteftines ; from which
                                        " inflam-

" inflammation arife intolerable pains, an acute
" inflammatory cholic, and afterwards either a
" gangrene and death, or a fcirrhus that dege-
" nerates into a cancer, (of which I have feen a
" dreadful inftance) or elfe by fuppurating,
" brings on an abfcefs or an open ulcer. Very
" often the matter is repelled elfewhere, and
" produces indurations in the liver, anxieties
" about the precordia, apoplexy, epilepfy, rheu-
" matic pains, fore eyes, and incurable cutane-
" ous diforders. Such, fays Dr. Tiffot, are the
" effects of all binding and narcotic medicines,
" as the theriaca, mithridate, and diafcordium,
" when they are given too early in this diftem-
" per. I was once called in to a very terrible
" rheumatic cafe, caufed by the theriaca be-
" ing given on the fecond day in the dyfentery."

Having premifed thus far from the obfervations of others, I now come to the main point,
namely, to what our own experience has taught
us in this epidemy, with refpect to binding, conftipating, and aftringent remedies. I could have
been furnifhed with an infinitely greater number of cafes from all parts, if people had not
carefully buried in filence thofe faults, againft
the commiffion of which they had been fo earneftly and urgently cautioned ; and if mankind
did not ufually ceafe to be fincere, as foon as
they perceive that they are obferved.

H

A young

A young skinner at Arau, put a stop to his dysentery by means of oatmeal-gruel, a remedy prescribed by the physicians of Bern in 1750, which he made very thick, and by the use of it, became lame hand and foot; in December he could neither walk nor work, his hands and feet were immoveable, and were perceived to wither away daily.

A man of forty years in the county of Lentsburg had the dysentery; a quack gave him a restringent medicine, his looseness disappeared, and immediately he fell into a rheumatism, that tortured him to the height of despair.

A country girl of eleven years, in the same county, got from a hangman * some restringent medicines, on the eleventh day of her disorder; immediately upon this her pains and looseness disappeared, and on the contrary her feet and abdomen were swelled up, and remained in this condition till her death, which followed in less than a month.

A peasant of thirty years of age, from Solothurn, had the dysentery; the same hangman gave him what he called his opiate drops, on taking which he began to swell, and lost the use

* Who in some parts of Germany, practise physic with great success; especially in desperate cases, when the patient is given up by the physicians.

of

of his hands and feet. Towards the end of December, he was carried about in Arau from one phyfician to another, imploring their affiftance.

Our peafants that were troubled with the dyfentery, often drank likewife warm milk. This feemingly innocent remedy, was in fome important cafes very noxious; their ftools were leffened by it, and even entirely ftopped; the patients were feized with a violent pain in the limbs, and a total relaxation and wearinefs, that made them unfit for all kinds of work.

At leaft in Thurgaw Dr. Keller perceived no peculiar good effect from milk, and ftill lefs from oil. To befure many people boafted, that they had cured themfelves by drinking plenty of milk warm from the cow. This happened without doubt, where there was only a purging to be cured, if at the fame time they ufed a good diet; but in a real dyfentery, Dr. Keller obferved very little benefit proceed from milk.

In the fame city Dr. Dummelin, town-clerk at Frawenfeld, faw two children, one ten and the other thirteen years old, who at the beginning of their dyfentery, had been made to drink plentifully of warm milk from the cow, directly upon which they complained of a painful contraction

and

and oppreſſion of the ſtomach, after which a
vomiting enſued, whereby the milk they had
taken, was brought up curdled into a cheeſy
matter moulded into the ſhape of dogs excrements; upon this convulſions enſuing, both the
children died in a few days. Dr. Dummelin
remarked the ſame bad effects of milk on children, in the epidemical dyſentery of 1738 and
1739.

This year all kinds of binding, conſtipating,
aſtringent, and ſoporific medicines, of all ſorts,
forms, and colours were made uſe of in Thurgaw. The moſt famous ſpecifics among the
common people, were red wine with pepper,
mutton baked in ſuet, brandy, and acorns
pounded and boiled in red wine; moſt of them
died, but ſome recovered. Acorns are indeed
adviſed by the Proviſion of 1750, (iſſued out by
the college of phyſicians at Bern) as a good remedy; on the contrary, the Council of Health
at the ſame place, has forbiden them as noxious,
in an edict read in the churches the very ſame
year. In ſhort, acorns are extremely aſtringent, and excite the moſt obſtinate conſtipa-
tions of the bowels.

Moſt of the country people of Thurgaw,
made uſe of cakes compoſed of mutton fat with
eggs

eggs and mint, under the denomination of approved family medicines. Many took powdered tormentil root, others lapis hæmatites, others gun-powder in a foft boiled egg, and others made ufe of garlick. Thofe that were only attacked with a purging, or had but a flight attack of the dyfentery, found no bad effects from thefe things. But in fevere fickneffes, a great wearinefs over the whole body was the confequence of fuch meafures; and the patient fell into a dropfy and cachexy.

The barber-furgeons of the villages in Thurgaw, began their cures for the moft part with aftringent medicines. In all their patients the malady was extremely heightened, and commonly death was the refult; fo that thefe beard-doctors were at laft forced to confefs, that the diforder was above their reach.

A practitioner of Thurgaw limited his whole method of cure to two remedies. The firft day he gave a mixture of ipecacuhana and rhubarb, the fecond Sydenham's liquid laudanum, and kept to this laft, till the purging remitted. On the firft of December, when I received this account from Thurgaw, all this practitioner's patients without exception were dead, either of a dropfy or a terrible rheumatifm, or elfe lay

eagerly

eagerly expecting the near approach of friendly death. In the mean while, the people of Thurgaw were too ſtupid to ſee into the deletoriouſneſs of the method purſued by this deſtroying angel, though manifeſted by ſuch flagrant proofs. Half Thurgaw cried out, that theſe people did not die of the dyſentery, but of the dropſy and rheumatiſm.

In the city of Solothurn, the uſe of aſtringents and aromatics, according to the obſervations of the excellent Dr. Gugger, increaſed the tormina and fever, and produced a gangrene in the bowels. But nothing had more fatal effects, and was more certainly mortal, than the improper application of laudanum.

To the effects of a dyſentery treated with reſtringents inſtead of good ſmart purges, I aſcribe the caſe of an Engliſhman of diſtinction, whom I reſcued from the moſt imminent danger by the means of purgatives and antiſeptics ; while at the ſame time, without doubt, many a timorous phyſician by leaving the dyſenteric matter in the body, would have left the patient in his diſorder.

This gentleman came into my neighbourhood from Florence, on the 7th of Auguſt.
                                                        He

He had been eleven days before attacked with a dyfentery, not far from the Baromean iflands. The Italian phyficians gave him indeed manna twice in the beginning, and the firft time a pretty ftrong dofe of it; but directly thereupon endeavoured to put a ftop to his flux, by means of opium, and other reftringent and binding medicines. The fick gentleman in the mean while haftened towards Switzerland with the dyfentery upon him, and taking thefe medicines; he performed his journey in the hotteft weather imaginable, on horfeback, as one is obliged to do in this country; the exercife of riding feemed to hearten him; he came luckily on the cloud-capt top of St. Gothard's, from the fultry air of Italy, into a piercing cold region. An Italian phyfician, whom he had taken to accompany him in his journey, gave him every evening a reftringent medicine. But nature was wifer than art; for the gentleman, after two uncommonly large ftools, which his phyfician undoubtedly did not mean he fhould have, and which he neverthelefs had on the fixth or feventh of Auguft, on his arrival at Zurich, found himfelf well. In this condition he came the fame day into our parts, where he intended to reft himfelf a little after the fatigue of his journey, and give himfelf up to my care for his perfect recovery.

H 4

I found

I found him perfectly hearty, without the least difagreeable fenfation in his abdomen, without the leaft irritation to a ftool, without fever, and not very weak. Notwithftanding this, as I thought the matter was not fufficiently evacuated, I therefore advifed him to take in the morning and evening, a large fpoonful of the tincture of rhubarb made with water, and at the fame time to fubject himfelf to a ftrict diet, fuch as was proper for his circumftances.

The eighth of Auguft he told me early in the morning, that he had had two good natural ftools, had refted perfectly well, and found himfelf entirely recovered. I ordered him however to go on with the rhubarb in the fame manner, till evening, and even at fupper he found himfelf perfectly well, talked chearfully without interruption, played at cards, and wrote.

The ninth of Auguft I was called to him early in the morning, in the greateft hafte. He told me that he had had two ftools in the night, which were not offenfive to the fmell, but pretty large; that he had not flept at all, had a fever on him, had been extremely reftlefs, and was ftill fo. I found fcarce any change in the pulfe, and contented myfelf with ordering him to take a fpoonful more of the tincture of

rhubarb

rhubarb the fame morning, and every two hours afterwards half a glafs of almond-emulfion, in the mean while obferving the courfe of the diforder.

In the afternoon I found the fick gentleman in a very difmal condition. His pulfe was really quicker, and he had a vaft heavinefs and pain in his head. Towards evening, an uninterrupted flumber for fome hours put an end to thofe fymptoms. At the beginning of the night he grew exceffively weak, and after that dofed till morning, while at the fame time the fever was very confiderable. It was however hard for me in this confufion to determine the fpecies of his fever. I therefore refolved to continue till the morning with the almond-emulfion, to fee if perhaps there was not fome matter in the body, which it might be neceffary to evacuate. His head-ach was diminifhed by the morning, but his pulfe was ftill pretty quick.

On the twelfth day of his diforder, on the mere prefumption, that fome of the dyfenteric matter might have ftill remained behind in his body, and have putrified by degrees, I gave him two ounces and a half of manna in water, with half an ounce of cream of tartar at one dofe. The effect of this medicine was, that an

aftonifhing

aftonifhing quantity of an extremely fœtid bilious matter came away, with a great number
of ftools, without the leaft fenfation of pain in
the abdomen, and with the greateft and moft
perfect alleviation, which further increafed at
every ftool. The fick gentleman found himfelf very hearty, and without the leaft indifpofition, till after one o'clock in the afternoon.

At two o'clock came on a violent cold fit,
that fhook his body all over, and continued in
this manner for the fpace of three hours, with
an unquenchable thirft, violent head-ach, and
no inconfiderable naufea. On this cold fit
followed a vaft dry heat over the whole body,
and an uncommonly violent fever, attended with
great terror and delirium. I now prefcribed an
ounce of cream of tartar, ordered it to be divided into twelve dofes, and one of them to be
taken every hour in an infufion of elder flowers;
at the fame time I advifed him to drink plentifully of lemonade.

The effect of this remedy, was an aftonifhing
copious excretion of a putrid and inconceivably
offenfive matter. At break of day I faw this
attack end in a ftinking fweat, which fmelled
like that of an intermittent fever.

The

The third day in the morning I found the gentleman perfectly well again. I prescribed once more a potion of two ounces and a half of manna, and half an ounce of cream of tartar. Again there came away an immeasurable quantity of a perfectly putrid matter. In the evening he found himself very well, and rested well the whole night. I now let him drink nothing else than a good quantity of lemonade.

The fourth day I went early in the morning and found him very chearful, hearty and well. I prescribed him an ounce of cream of tartar to be divided into twelve doses, one of which was to be taken every two hours with lemonade, about noon I was sent for on a sudden. Another equally violent cold fit, shaking the very inmost parts of the body, began at half past ten, and kept on in the most violent manner till one o'clock, during which the patient very frequently vomited, and went to stool. After the cold paroxysm, ensued the same violent parching fit and unquenchable thirst, high fever and delirium, as before. However, the attack seemed inclinable to cease at ten in the evening, and soon after actually did so. I let him go on with the cream of tartar and limonade, according as I had ordered in the morning.

In

In the night time he refted tolerably well, at leaft as to his body.

The fifth day in the morning I gave a vomit of half a drachm of ipcacuanha, which worked very eafy, but did not difcover any more than before, any unufual matter harbouring in the ftomach. During the whole morning there ftill enfued a great number of very offenfive ftools, by which the fick gentleman was very much relieved. In the attack of the fourth day I had remarked, that during this unea'nefs of mind, which proceeded principally from the nature of the diforder, the whites of his eyes were intenfely yellow. I feared therefore, that at every future attack, the bile would flow in great quantity into the inteftines, or return back into the blood ; in fine, that from a fimple putrid tertian fever, it might turn to a double putrid tertian ; and confequently the diforder of this gentleman, who was of fo much confequence to his country, might grow more and more difficult and dangerous. On thefe confiderations I refolved now to make ufe of the barĸ, and to give a full ounce of it before the next attack, which according to the premature courfe of the diforder, might be expected on the fourteenth of Auguft, at fix in the morning. At two in the afternoon I made a beginning,

ginning, and by four o'clock in the morning
the whole ounce was confumed. His ftomach
rofe againft the bark, and a ftrong irritation to
vomit enfued, and with thefe his ufual anxiety
of mind. I had it in my power to put a ftop
to his vomiting, but I let his ftools, which
were pretty frequent, take their courfe; as I
did not look upon them in a bad light, but
merely as the effects of the bark.

In the evening and beginning of the night
the pulfe was unequal and fluttering, and at
times quick; however, this I attributed to
the condition of his mind.

On the fixth day his pulfe was juft as it
fhould be, from early in the morning till nine
o'clock, and the gentleman himfelf very chearful. After ten he had a little fenfation of cold
in his body, but nothing like a regular fhivering fit, though his hands were at the fame time
perfectly warm. After eleven o'clock his dejection of mind came on, and upon that a
middling heat, that continually increafed, and
towards evening became very confiderable, being
attended with much fever, and an uncommon depreffion of fpirits; after eight o'clock this attack,
already moderated by the bark, had an end.
The whole day throughout I had given him

no medicine at all. I now prefcribed an ounce
of bark, ordered it to be divided into fixteen
dofes, and one of them to be taken every two
hours, beginning from ten in the evening;
whence at each time during the whole night
enfued a very offenfive ftool, but no irritation
to vomit.

On the feventh day the fever came to a perfect intermiffion, though the patient was not
free from a dejection of fpirits very natural in
fuch circumftances, which towards evening
bordered upon abfolute melancholy. Till eleven
at noon a ftool followed upon each dofe of
the bark, which he now took every hour. His
urine, which fince the accefs of the fever, had
come away in ftreams, was ftill bloody. He
paffed the night fleeplefs, but entirely without fever.

On the eighth day * he was perfectly chearful
all the morning long, the brighteft ideas fhone
in great profufion through his whole difcourfe,
and his departure was abfolutely fixt for the
next morning.

* The fever ought to have come again, but not the
leaft glimpfe of it appeared during the whole day.

On

On the ninth day he fet off on his journey.
I gave him another ounce of the bark, which
he was to confume in the fpace of that day, the
feventeenth of Auguft; I advifed him to take
an ounce more on the nineteenth; and a week
afterwards, and fo on, to repeat the fame quantity, in order to prevent a relapfe. I cautioned
him very ferioufly not to take any purgative
medicine within the fpace of a month, if he
had not a mind that his fever fhould directly
return. I advifed him in the mean time to live
entirely upon vegetables.

The fecond of September 1765, I had an account from fome diftance off, that this gentleman found himfelf quite well on the twenty-
fourth of Auguft. But the phyfician, whom
he had at that time confulted, had thought it
neceffary to add rhubarb to the bark, on which
his fever had returned directly the very fame
day; however, it was at that time cured by
another phyfician. On the fixteenth of September the Englifh gentleman wrote me word himfelf, that he now perceived no difference between
one day and another, and was in perfect health.
And indeed, fince this time he has always had
the beft health in the world; and enjoys at
prefent, in the fixty-fourth year of his age, fuch
alacrity and ftrength of mind, that he actually
                                        tranfacts

tranfacts the moft important and perplexed'
ftate affairs, with unfpeakable facility and great
reputation.

After this long but ufeful digreffion, I refume
the thread of my difcourfe, and haften to relate
the noxious effects of aromatics, brandy, and
wine in our dyfentery. ·

Aromatics and wine in general, excite in dyfenteric perfons a dangerous irritation in the
bowels, increafe the pain, fever, and ftrangury ;
and when they operate as aftringents, (which
however does not often happen) they produce
all the above-related bad effects of thofe dangerous medicines ; change the bloody excretions
into a thin pus ; and wine in particular excites
a very dangerous fenfation of anxiety in the pit
of the ftomach, that very often accompanies an
inflammation of the bowels, or precedes it or a
gangrene ; but which ought not to be confounded with that oppreffion of fpirits, that appears directly at the beginning of malignant dy-
fenterics. Brandy is here an abfolute poifon, and
all thefe remedies occafion, even in convalefcents,
a moft dangerous relapfe of the fame diforder.
All the Swifs phyficians, that find the peafants,
who are attacked with the dyfentery, in a very
perplexed condition, have the blame to lay on
the

the ufe of binding medicines, and their beloved kitchen phyfic ; but chiefly to nutmegs, mace, ginger, pepper, wine and brandy, which to be fure keep the dyfentery under for a while, but foon afterwards throw the patient into the moft dangerous diftempers. Dr. Tiffot faw once eleven fick of the dyfentery in one houfe. Nine eat fruit, and were happily cured ; the grand-mother and a favourite child of her's went to their graves ; for fhe took care of the child in her own way with mulled wine, oil, and fpices, and accordingly it died. She treated herfelf in the fame manner, and died in her own way too.

Dr. Tiffot faw in like manner a man troubled with the dyfentery, who had drunk about two ounces of brandy, fuddenly feized with a hic-cough, which he oppofed by the external and internal application of anifeed-water. Upon this enfued an inflammation in the ftomach, which almoft brought him to his laft·gafp ; but Tiffot was his phyfician. However he was fickly for about a year afterwards, but at length Dr. Tiffot made him well.

Now all thefe noxious remedies, together with putrid cheefe, were ufed in great quantities by the country people in our dyfentery : and

alfo

alfo by thofe in the cities, that lived after the country manner. Our peafants took immediately at the firft attack of the diforder, nutmegs, pepper and cheefe; in a flight indifpofition they got over it; in fevere cafes the vomiting continued, the medicines, that were afterwards given them, did not ftay on their ftomachs, and thus the patients were brought to their graves. In the county of Lenzfburg the country people, in obedience to the advice of the famous Senertus in the foregoing century, which is ftill propagated in our times by ignorant phyficians, univerfally made ufe of red wine and rotten cheefe in the beginning of the diforder. Upon this the price of cheefe evidently rofe; and an innkeeper near Solothurn, not far from Arau, fold during the dyfentery's raging in the canton of Bern, feven hogfheads of red Alfatian wine more than the ufual quantity he otherwife confumed in the fame number of weeks; accordingly, directly at the beginning of the epidemy in the county of Lenzfburg, a great number of people gave up the ghoft merely from this intemperate ufe of wine and cheefe. In like manner the mortality that happened in Thurgau at the beginning of the epidemy, was moft confiderable among the great number of fick, that lived in the whole extent of the fouthern fide of Ottenberg, as thefe people notwithftanding all

the

the admonitions that were given them, continually followed their depraved appetites with regard to wine and brandy. At laft fuch of the people of Thurgau, as ftill remained alive, and were
not yet feized, grew fomething wifer by the
conftant tolling of the bell. They purfued a
better diet, and rather had recourfe to the phyficians, than to the wine-cafk or cherry brandy-
bottle.

But this is not fufficient for me; for I pro-
pofe to relate alfo fome fpecial obfervations
made in our epidemy, on the dreadful effects of
aromatics, brandy, wine, and other things of
this nature.

A young woman of twenty years of age in
Brugg, had had the dyfentery eleven days to a
very high degree. She was under the care of a
very fenfible phyfician, who did me the honour
on the eleventh day to call me to the confulta-
tion. The evening before the patient at the infti-
gation of an old woman, very famous amongft
us for her fkill in phyfic, had drank a good deal
of wine. On this followed the whole night
long, vaft pains in the abdomen, copious ftools
full of blood, a violent fever, delirium, and
cold fweats. I found too at the time that I was
called, her ftools very numerous, painful, and

bloody;

bloody; her pulfe extremely quick, and the
patient herfelf full of anxiety, and in particular
a violent preffure in the region of the fcrobicu-
lum cordis, which Morgagni fays, is not feldom
the immediate forerunner of death. I did not
dare to think of any evacuation; for all things
confidered, I concluded that the wine had already abfolutely occafioned an inflammation,
for which reafon I ordered nothing elfe, than
every two hours two fpoonfuls of a mixture
compofed of half an ounce of gum arabic, four
ounces of water, and one ounce of fyrup of
marfh-mallows; and with it a great quantity of
almond-emulfion, rice-gruel, gum arabic clyfters, and the fomentations commonly ufed in
inflammations, to be applied to the abdomen.
Towards evening fhe had again a fhivering fit,
but no delirium in the night. The twelfth day
I found her ftools diminifhed in number, and of
a green colour; fhe complained perpetually of
a fenfe of heat in the pit of the ftomach. I let
her ftill go on with the fame medicines, but was
defired to omit the clyfters. During the whole
day our patient feemed very much relieved, but
in the evening her pains, and particularly the
tenefmus got again the upper hand. I advifed
them to continue ufing the fame medicines without interruption, and in the night time to give
her two gum arabic clyfters. The next day
all

all the symptoms were very much diminished, and in a few days she was perfectly recovered.

A farmer's lad of thirteen years of age, in the district of Wildenstein, was taken with the dysentery. He had the affistance of a very famous physician at Brugg, Dr. Fuchslin, and was happily cured with purgatives. But the seventh day he drank wine, and eat a pretty deal of cheese; the dysentery returned with violent tormina, and a pretty considerable lofs of blood; Dr. Fuchslin cured him once more. A week afterwards he drank wine again, and again eat as much cheese as he could ftuff; the dysentery returned with still greater violence, and continued a month.

A peasant eighty years old in the same district, was attacked with the same diforder. He made a shift to crawl as far as Brugg; however not to a physician, but to a wine cellar, devoured a good quantity of cheese, washed it down with a quart of red wine, staggered home, went to bed in the moft dreadful pain, got upon that a bottle of wine from the good-natured parson of his parish, sent for advice on the tenth day of his diforder to a quackfalver in the county of Baden, and on the thirteenth day died.

A very

A very hearty and otherwife healthy farmer's lad, fifteen years old, in this diftrict, was attacked with the dyfentery in fuch a flight manner, that he was able to go about with it during a whole week. His mother undertook his cure; gave him a mixture of red wine, cheefe, nutmeg and pepper, and by the fourteenth day he was dead.

Another of fixteen years, in the fame diftrict, was feized in the fields with a fhivering. The next day he went again to field, complained, when he was there, of a great wearninefs in his members, and directly upon it laid himfelf ftretched out for half an hour, during a violent fhower of rain on the wet earth. On the third day he had a perfect dyfentery with great bodily pains; on the fourth a violent vomiting. On the fifth I was confulted; I gave the ufual remedies; but he took only the emetic, and that with much relief; the reft of the medicines he threw away and drank wine in their ftead. The eighth day out of compaffion I went myfelf to this fellow, and tried to prevail on him in the moft friendly, genteel, and perfuafive manner; but I loft my labour. He did not defire my affiftance, for he had at that time medicines in the houfe from a

quack-

quackfalver in the county of Baden, and withal
a great inflammation in his bowels, of which he
died the next day.

A country girl of eighteen years of age, in
this diftrict, was feized with the dyfentery. I
was applied to for advice, but my patient did
not take half the medicines I had ordered for
her; but during the time fhe fhould have taken
them, partly made ufe of an elixir which fhe got
from the charitable parfon of her parifh, partly
of an elixir that her ftupid mother (who was
the oracle of the village) gave her repeatedly in
wine. With this the mother made her daughter follow a moft abfurd diet; fhe put wine in
all her broths, gave her flefh, curds, and paf-
tery, that an oftrich would fcarcely be able to
digeft; at laft fhe threw away all her medicines,
fo that the dyfenteric matter remained behind,
notwithftanding the infignificant excretions
caufed by the diforder, and the putrid fever by
thefe means was ftill kept up. On the twenty·firft
day of the diforder a miliary eruption, ·and a
confiderable abfcefs in her body made their appearance; the dyfentery continued, and her
ftools remained as before, yellow, green, brown,
red, and black. They fent for the prieft,
whined, prayed, and made ufe of fuperftitious
remedies, hanging a piece of fcarlet cloth
I 4 about

about the patient, in hopes of banishing the imaginary scarlet fever.   When all this was of no service, the girl's father asked advice again of his priest, who told him that a sick person, whom I had forbid wine, had drank two bottles, and was perfectly cured.   Upon this the father came home like a mad man, and cried with many gesticulations, that his daughter should not only have as much wine as she chose, but every thing else she pleased.   All this was accordingly done.   However, when on the twenty-sixth day of the disorder, all this had been of no benefit, her father applied to me once more for advice.   Quite astonished at the honour he did me, I endeavoured with a friendly earnestness to open his eyes to all this folly, and touched his conscience in such a manner, that he promised me with tears in his eyes never to listen to a woman any more.   I ordered him therefore to give her some doses of cream of tartar, and three ounces of tamarinds dissolved in water. These medicines brought away a good deal of the usual matter with much alleviation, after this her stools decreased, her appetite increased, and the miliary pustles went off by desquamation.   On the twenty-eighth day of her disorder, her father told me that his wife had given her in the morning a pretty deal of mulled wine, upon which she found herself again very ill. The

The good fuccefs of the tamarinds I lately gave her, encouraged me to try them again once more in this extreme danger, which I pointed out to the father; my patient took the tamarinds, and at the fame time her mother gave her butter-milk, whey, muft, and every thing that came into her head. After that they · afked my advice, as I rode by the door, but I gave them no anfwer. On the thirty-fourth day of the malady the patient died.

A wholefome, hearty, clever farmer's wife of eighteen years of age, in this diftrict, was flightly feized with the dyfentery in the third month of her pregnancy. Her mother gave her directly upon it every day three glaffes of brandy, and at the fame time white and red wine in great quantities. The firft effect of this was her mifcarriage on the third day, which was followed by a very great lofs of blood. They went on boldly with the brandy, her legs grew cold, fhe had a mortification in her bowels, and died the fifth day.

A woman of feventy-eight years of age was taken with the dyfentery in Arau, and made ufe of the ufual fpecific of nutmeg and red wine. The fecond day of her diftemper Dr. Seiler, a phyfician of great ingenuity and penetration, was

fent

fent for, who found all the fymptoms of a violent inflammation in the bowels, but notwithftanding all he could fay, the woman took a
whole deluge of her fpecific, and died in four
days.

A country girl aged fifteen years, in the
county of Lenzfburg, lay fick of the dyfentery;
they gave her red wine in the firft days of her
diforder ; on the fixth day enfued a hiccough,
and on the fourteenth, death.

A batchelor thirty-fix years old, in the fame
county, drank red wine in the fecond day of
his dyfentery. He fell immediately into a delirium, and the fifth day he was feized with a
continual hiccough, and on the fourteenth by
death.

. A man forty years old, of the canton of
Zurich, was vioiently attacked with the dyfentery;
accordingly he had immediate recourfe to the
fpecific qualities of red wine feafoned with nutmeg. Neither copious bleeding, nor ipecacuanha,
which was given afterwards on account of his
belching and naufea, nor cooling purges, nor
any other remedies that were given him by an
excellent phyfician in Zurich, were capable of
abating the violence of this dyfentery, and the
in-

intolerable pains attending it. The quantity of blood that came away was fupernaturally great, and fo early as the fecond day the phyfician thought he faw in the matter that was voided, the figns of a wound in the internal coat of the inteftines ; on the ninth day the patient died.

Under the denomination of approved family medicines, the country people in Thurgau (according to the account of Dr. Dummelin at Frawenfield) made ufe not only of old red wine in different forms and compofitions, but even applied to melaffes, genuine brandy, fpirits of blackberries and juniper ; fuch as were feized only with a diarrhœa, or even a flight dyfentery got over it ; but with moft people the wine and fpirituous liquors increafed the gripes, ftools, and tenefmus, caufed a great fenfation of ardor in the ftomach and inteftines, heightened the fever, occafioned great heat, intolerable thirft, vaft horrors, and at length death itfelf.

A man in Frawenfield, that was almoft got out of danger by the affiftance of Dr. Dummelin, produced fuch an alteration for the worfe in his dyfentery, by taking a good draught of wine, that a fingultus enfued, and upon that a vomiting of blood, and finally death.

Many

Many people in Thurgau oppofed the firft fymptoms that appeared of the dyfentery, by drinking plentifully of a mixture compofed of red wine and fpices ; their loofenefs was ftopped, and they began to triumph. After fome days more or lefs, Dr. Keller of Weinfield, faw fome of thefe fall into a relapfe, which was worfe than the firft attack ; others were in a moft painful and miferable condition, which (as Dr. Keller exprefles it) was a perfect affemblage of all the evils that plague mankind. At the leaft touch of any thing befides the bed-cloaths, they fuffered in the part offended fuch violent pains, that they broke out into dreadful howls, and gave the utmoft tokens of defpair. Dr. Keller had two of thefe wretched victims to popular felf-conceit under his care ; and cured them both with copious bleeding, and the long-continued ufe of antiphlogiftics.

The fick in Thurgau to the north of Ottenburg, who did not follow the moft ftrict diet, and in particular did not entirely abftain from brandy, wine, and flefh, died almoft all of them between the ninth and twelfth day.

In Swabia too, according to the obfervations of Dr. Moehrlin, nothing was more noxious and dangerous in the dyfentery than wine, except
brandy,

brandy, which was ftill more fo. Thofe (faid
Dr. Moehrlin) that in the courfe of their illnefs
drank wine, could not be faved. Such, as juft
before their being attacked with the diforder,
made ufe of wine and brandy by way of prefer-
vatives, had it uncommonly violent and long,
and were alfo at the end of it troubled for many
weeks after with obftinate œdematous fwellings.

A woman in Ravenfburg had put a ftop to
her dyfentery by the copious ufe of red wine;
the effect of this was a great wearinefs, a pun-
gent and excruciating pain in one leg, at length
a perfect arthritis (which rendered her incapable
of moving in her bed) and a great difficulty of
breathing.

At length a cafe very appofite to illuftrate
this fubject came by chance into my hands,
from which I immediately made the following
extract, and which deferves a place in the very
important doctrine of the gradation of diforders,
as well as the above-mentioned cafe of the illuftrious Englifhman, which fell under my own
obfervation.

A lady of diftinction in Swabia was on the
eleventh of June, 1765, attacked on a fudden
with

with à loofenefs, which was at length attended with gripings and a bearing down in the rectum. On the fifth day fhe took of her own head a dofe of Englifh falts ; this in the opinion of the phyfician, who is the author of this account, produced a very bad effect, as her ftools became bilious upon it ; fomebody gave her befides this the fame day, eighteen drops of a diftilled oil.

The fixth day the phyfician of the place was called. He found her in the condition juft defcribed, that is to fay with the dyfentery on her ; and gave her a drachm of powder of rhubarb, which worked her pretty ftrongly, and tinged her ftools with blood. Without any longer delay he proceeded directly with a powder compofed of coral, cryftallus montana, cornu cervi uftum, dragon's blood, and cafcarilla.

On the eighth day another phyfician was called to his affiftance. The patient had ftill fome twitchings in the bowels, and her ftools were mixt with blood ; her two phyficians gave her a powder of gum arabic mixt with a little ca[...]illa, to be taken in an almond-emulfion or barley-water.

On

On the ninth day the patient had in the morning a pretty natural stool, but again in the evening a dysenteric evacuation with gripings and pain in the upper part of the fundament. Both the physicians upon this added to the powder the incomparable theriaca. On the tenth day she had some few pretty good stools, without any bearing down, though they were still mixt with pure blood; the pain about the os sacrum still remained, and her pulse was natural; her physicians now mixt with the abovementioned powder, instead of the cascarilla, two grains of jesuits bark. On the eleventh day her stools were very good, upon which her physicians gave her every five hours ten grains of the cortex without any additition, in order to strengthen her stomach and bowels; in the afternoon some hysteric symptoms appeared, upon which account they added to each dose of the bark, one grain of extract of castor.

On the twelfth day she was in the same condition, and towards evening she had her menses. The physicians ordered her to lay aside her medicines; on the other hand, they indulged her with two spoonfuls of Burgundy wine every six hours, upon her earnestly asserting, that when ever she was attacked with hysterics, even tho' they were attended by a fever, a glass of Burgundy

gundy was her beſt remedy.   They found upon
this with great pleaſure, that her pulſe roſe after
taking this wine, for which reaſon they allowed
her on the fourteenth day to take two ſpoonfuls
of Burgundy every four hours, and one of
them took his leave.

On the ſame night the phyſician who was retained, was informed, that the lady was again
ſeized with the hyſterics, upon which he ſent
her directly a grain of extractum croci.   On
the fifteenth day, early in the morning, the
phyſician went himſelf ; he found her in great
terrors, with ſpaſms, reſtleſſneſs, heat, great
thirſt, and a quick, violent, irregular pulſe. This
ſudden attack of a manifeſtly febrile diſorder,
the phyſician miſtook for the effect of a fright
ſhe had had the preceding evening ; in conſequence of which he gave her only ſome remedies
for her fright, and in particular the pulvis
margravii with the extract of caſtor.

On the ſixteenth day the other phyſician returned.   The lady lay ſince the morning in violent, and almoſt intolerable terrors, and
complained moſt of a very great anxiety and
weight at her breaſt ; the fever, with all its
ſymptoms, was more violent than the day before.   Her phyſicians gave her again nervine
remedies ;

remedies; namely, two grains of the pulvis margravii, with a grain of extract of caftor, to which they now added two grains of nitre. In the evening the lady had two ftools, in order to prevent any more of which, her phyficians gave her inftead of the pulvis margravii, an addition of coral to the powder compofed of the extract of caftor and nitre, to be taken every four hours.

The feventeenth day early, the lady had convulfive twitchings in her right arm. Thefe increafed ftill more and more, and came by degrees into the left arm, at length to her head, where they manifefted themfelves by contorfions of the eyes, noife in the ears, and diftorfions of the mouth and countenance; her eyes grew red, muddy, and dim, her face bloated and livid, and her reafon troubled. The phyficians thought of taking away fix ounces of blood, upon which every thing was on a fudden quieted. In the afternoon the patient had only now and then a depreffion of fpirits, which however foon vanifhed. The whole night throughout fhe was pretty quiet.

The eighteenth day in the morning the lady had a very profufe and extremely fœtid fweat,

K   which

which however went off by reafon of their having taken the precaution to change the fheets.
Upon this followed infupportable terrors, with
violent convulfions, and fpafmodic refpiration;
her eyes were ftaring and diftorted; her thirft
unquenchable, and her pulfe tremulous; her
phyficians tried in vain to recall the fweat; with
this view they took away four or five ounces of
blood, upon which the fymptoms remitted, but
did not abfolutely ceafe. After this they gave
her nothing but fudorifics; the fymptoms at
firft increafed, but at length difappeared about
noon, when fhe had a fmall evacuation; in the
afternoon the fame fcene was going to be
played over again; but upon the fudorifics being repeated, the fymptoms difappeared, and
at the fame time the fweat. In the evening
the lady complained of a violent wandering
pricking pain in her breaft and abdomen, and
faid fhe found fomething in many places rolling
itfelf up like a ball; the pain in her belly difappeared on the application of warm napkins,
but remained the whole night in her breaft.

The nineteenth day early fhe broke out again
into a fweat, by means of which the pains of
the abdomen went off. At firft the patient
appeared pretty quiet; but by degrees they
remarked a diftorfion in her eyes and countenance,

nance, her mufcles feemed convulfed, and her arms contracted. The lady faid, that at times fhe faw fomething frightful, and fomething was obferved in her converfation that was not natural to her. Both her phyficians upon this took it into their heads, probably on account of their apprehending fome malignity in the dif-order, to apply a warm hen cut up alive to the foles of her feet, and another to her head, which was accordingly done. In half a quarter of an hour the lady was quieter, and the phyficians took courage again; as fhe faid, that from the application of the hen to her head and feet, fhe felt an agreeable fenfation in the back. In the mean time about noon, a delirium enfued, and foon after that a fhort flumber, out of which fhe never awoke.

To this remarkable hiftory the following ac-count was added by one of the deceafed lady's phyficians. In her firft illnefs, which was manifeftly a dyfentery, her pulfe was never ex-traordinarily feverifh, and when fhe was free from hyfterical attacks, quite natural. Thefe attacks were repelled by the ufe of foot-baths, anifeed, cinnamon, mint, and chamomile-wa-ters; in her fecond illnefs her pulfe was as ir-regular as poffible, though always quick, with

K 2   a vio-

a violent and continual heat; her urine in fmall
quantities, thick, of an extremely high red
colour, and without fediment : her ftools were
liquid, of a whitifh yellow colour, and at
times covered with froth.

Both her phyficians joined in roundly declar-
ing, that they were not afhamed to fubmit the
hiftory of this diforder to the judgment of every
fenfible man; upon an impartial refearch they
muft neceffarily be allowed to have done right ;
that the phyfician cannot always cure, and that
the diftemper is fometimes too much for his art,
though exercifed with the greateft judgment.
In fine, they both concluded in a manner which
I did not expect, viz. what external preceding
caufes have occafioned the dyfentery to turn
to a violent convulfive diftemper, is notwithftanding all our refearches to us ftill hidden and
enigmatical.

With thefe learned gentlemen's leave, the
evident, true, and only external caufe of this fatal
change of the diftemper into an acute one was,
the Burgundy wine; and the evident, true, and
only internal caufe of this change, was the not
fufficiently evacuated, though difficultly retained
dyfenteric matter.

From

From this collection of obfervations and experiments it is clear, that aftringents and conftipating medicines, fpices, brandy, and wine, were very noxious, and always extremely dangerous in our dyfentery; and that it would perhaps be worth while to weigh in the fcale of humanity that political maxim, " Where no " accufer is, there is no judge."

CHAP. VII.

Remarks, Obfervations, and more determinate Conclufions, with regard to the Diagnofis and Cure of moft Species of the Dyfentery.

A STRICT enquiry into, and precife determination of the different kinds and fpecies of the dyfentery, which next to the plague and peftilential diforders, is one of the moft dangerous, and at the fame time, one of the moft common diftempers that infeft mankind, muft, I think, neceffarily have a great influence on the conduct of the phyfician, that intends to cure it. An effential diftinction is ufually made between a dyfentery with fever and one without fever, a benignant and malignant, a contagious and not contagious dyfentery. Without the moft abfo-

lute

lute and precife infight into the truth or falfe-
hood of thefe diftinctions, it is impoffible to
have a good fyftem in one's head; and without
a genius for practice, in intricate and irregular
cafes, a man appears like a fool at the bed-
fide of his patient, with all the fyftems in the
world.

The effential diftinction that is made between
a true dyfentery with fever, and one without
fever, appears to me to have a very dangerous
tendency, and, in my opinion, fhould be banifhed from the fchools of medicine; as it rather
determines the limits between a dyfentery and a
diarrhœa. To be fure, the fever that accompanies the dyfentery, feems very often at firft
almoft inconfiderable; upon which account
many phyficians affirm, that the dyfentery is fo
far from being attended with a fever, that its
concomitant fymptoms are of an almoft oppofite nature, viz. a pale face, the pulfe not
quicker than ufual, and fmall, and the diforder
of long continuance. But chillnefs, weaknefs,
and lofs of ftrength on the firft attack, are
however the ufual forerunners of a genuine fever, and thefe are never wanting in the regular
attacks of a true dyfentery. It is true, that the
pulfe in the firft days of the difeafe, is very often
not quicker than common, and very fmall; but

it

it appears to be manifeftly quicker in the courfe
of the diforder, and often extremely fo. I even
faw fome cafes in a dyfentery of the putrid kind,
in the epidemy of 1766, commence with a very
high fever, and terminate very happily; while
others began with a fever that could hardly be
perceived, which however ended in death. In
the firft cafe the patient's countenance was as red
as fire, in the latter, pale. I find this fever of
ftill more confequence, when all the other fymptoms are bad, when the patient lofes all his
ftrength, and the pulfe is at the fame time not
quick, but extremely weak; for in this cafe
there is an undermining malignant fever, or
even a gangrene lurking behind the curtain.
In thofe that lie fick of the dyfentery, the fever
fometimes, and that even on the days immediately preceding the death of the patient, feems
to be quite vanifhed, while, at the very fame
time, the inflammation is haftening on to a gangrene. Upon this account, the effects of the
dyfentery have been very aptly compared to the
effects of arfenic; as this, juft like the dyfenteric matter occafions reachings, copious ftools
which corrode the bowels, pains at the heart,
terrors, intolerable belly-achs, inflammations,
gangrenes, and death, without any confiderable
fever being at the fame time to be obferved. In
fine, the length of the malady is no argument

K 4   in

in this cafe, as the worſt fevers in ſome particular circumſtances viſibly run out to a great
length ; eſpecially the fever with us, ſometimes
very improperly, called putrid. Though theſe
obſervations made at the bed-ſide of the patient,
are not very agreeable to the greateſt part of
our ſyſtemwrights, they are however of the
greateſt conſequence ; as many pre-occupied
with a notion, that in a dyſentery of this ſort
the pulſe muſt be neceſſarily very quick, think
it accordingly ſlight, and of no ſignification,
when in this point of view the fever is imperceptible. This error was, according to Degner's account, very dangerous to many perſons
in the dyſentery at Nimeguen, in 1736; and
upon that account, I do not ſee why Dr. Aken-
ſide appeals directly to that epidemy, in order
to prove, that the dyſentery is not attended with
a fever.

Sydenham calls the dyſentery a fever turned
upon the bowels. This mode of expreſſion, I
muſt own, does not perfectly pleaſe me, as it is
not taken from the conſideration of the principal
phænomenon : this diſtinction, however, ſeems
to comprehend the very eſſence of a genuine dyſentery ; and I am convinced that this diſorder
ſhould be treated ſometimes as an inflammatory, ſometimes as a bilious or putrid fever,
ſome-

fometimes as a fever compounded of both, fometimes as a malignant fever, and fometimes as a bilious one, accompanied with manifeft tokens of malignity; but I am fenfible too that there are many cafes, in which it is equal how they are treated. I confidered this year's dyfentery in all ferious cafes, as a bilious or putrid fever; for I fhould have been much miftaken, if I had looked on it as a mere inflammatory cafe, and accordingly breathed a vein, and given neither vomits nor purgatives; or, if I had taken it for a malignant diforder, and accordingly allowed my patient wine, and ftrengthening remedies. Some phyficians were, to be fure, very liberal in letting blood in our dyfentery at Thurgau; as they (perhaps induced by an hypothefis) prefuppofed an inflammatory ftate of the blood in all their patients, and indeed might very naturally imagine this to be the cafe, on account of the propenfity of our countrymen to drinking. It may be, that in that part of the country, the dyfentery in many perfons was a complication of inflammation and putridity together; at leaft it happened to be fo with us, towards the end of fuch diforders as proved mortal; and indeed, as I imagine, is generally the cafe in thefe circumftances, juft before the diforder terminates in death, excepting where there are other peculiar caufes fubfifting for

the

the patient's death, totally different from this.
We ſhould however be very careful to diſtin-
guiſh, whether this complication ſhews itſelf at
the beginning of the malady, or whether it
only accompanies the fatal turn of the diſtemper. It is not impoſſible, that it ſhould happen
in the beginning of the ſickneſs ; and it was
obſerved to do ſo in our putrid pleuriſy, to which
an inflammation of the lungs was often joined
even at the beginning, that made emetics, which
would otherwiſe have been of ſo much ſervice,
prove fatal to the patient. In like manner that
kind of pleuriſy, which entirely belongs to the
bilious, or (as it is called) putrid claſs, and all
other ſimple putrid fevers often, like the dyſentery, terminate juſt before the mortal criſis,
in an inflammation and gangrene of thoſe parts
on which the putrid matter has ſettled. But I
alſo hold it as a very important maxim in the
exerciſe of our art, and perhaps not ſufficiently
taken into conſideration even by ſome great
phyſicians, that we ſhould be very careful not
to confound the divers periods of a diſorder one
with another, nor to take from what happens
towards the end of the ſickneſs, and ſtill
more eſpecially from the dead body, the indications of what is to be done at the beginning.

There

There are many circumſtances, which occaſion the junction of the dyſentery with a malignant fever; but eſpecially, when many ſick people lie together in a narrow place, when the chambers are neglected to be aired, and in general, when little or nothing is done, that the rules of our art require to be done in ſuch caſes. This fever may attack people in health, without being attended with the dyſentery, though it ariſe from the putrid and confined vapours of that diſtemper; it may however ſupervene on the dyſentery, and in this caſe it will be of a perfectly peſtilential nature.

After the battle of Dettingen, the dyſentery immediately made its appearance in the Engliſh army, and committed great ravages in it during the month of July and part of Auguſt. The hoſpital was in the village of Fechenheim, about a league from the camp; during the time that the army lay near Hanau, about fifteen hundred ſick, beſides thoſe wounded in the field of battle, were brought from the camp into this hoſpital, and among theſe the greateſt part was ill of the dyſentery; the air was by theſe means infected to ſuch a degree, that not only the reſt of the patients, but even the apothecary, nurſes, and the other ſervants, with moſt of the inhabitants of the village were infected.

fected. To this was added a ftill more alarming diftemper, namely, the jail or hofpital-fever, the common effects of a foul air, proceeding from a great throng of people, and animal corruption. Both thefe together occafioned in fo fhort a time as the month of July and part of Auguft, a great mortality in the village : while on the other hand, thofe that were attacked with the dyfentery, but were not removed from the camp into the hofpital, remained free from this malignant difeafe, and were happily cured; although they were in want of many conveniencies, which the others in the hofpital enjoyed. After the Englifh army had marched away for the Netherlands, in September 1743, three thoufand fick were left in Germany, part in this fame village of Fechenheim, and the reft at Ofthofen and Bechtheim, two villages in the neighbourhood of Worms. At Fechenheim the malignant fever and dyfentery grew worfe and worfe every day; few kept free from it, for whether the dyfentery, on account of which the fick were fent to the hofpital, was mild or malignant, this fever was always joined with it. The petechial fpots, blotches, parotids, frequent mortifications, contagion, and the great mortality, fufficiently fhewed its peftilential nature. In this point of view it was ftill worfe than the plague itfelf, as here there was always reafon

reafon to fear a relapfe; and it might almoft be depended upon, if the patient remained in the fame contagious air. Out of fourteen mates that were employed about the fick, five died; and all the reft, except one or two, had been ill and in danger. The hofpital loft near half of the patients; but the inhabitants of the village, by the dyfentery and fever together, were almoft entirely deftroyed. Now I conclude from all I have faid, and efpecially from the obfervations of that great phyfician Dr. Pringle, that not only a fever of different kinds, for the moft part accompanies the dyfentery, but that this fever in certain circumftances, is of the moft alarming nature.

It is not without reafon, that a diftinction is made between a malignant and benignant dyfentery; but at the fame time, this diftinction is little underftood, and occafions a great many miftakes; as a diforder is often called malignant, which in reality is not fo; and malignancy is often leaft fufpected, where flily undermining the conftitution, by flow advances it gets at laft the maftery, and overturns every thing.

It is undeniable, that many dyfenteries are of a good, and others of a bad fpecies; that

fome

fome only attack one here and there, while others at once, with their peftilential miafmata, infect the whole country. There is a kind of dyfentery peculiar to the ifland of Java, in the Eaft Indies, which is commonly of a benignant nature. Its beginning and progrefs is very flow, the ftools are not very copious, the pains not great, nor the weaknefs confiderable. A flight, and often imperceptible fhivering, which does not eafily return in the courfe of the malady, without fome error committed in diet, ufhers in the diforder. The ftools are liquid, yet not copious, infomuch, that fuch as are attacked with it, do not ceafe to tranfact their ufual bufinefs, and commonly do not apply to the phyfician for advice, before three or four weeks are at an end. By degrees the ftools become more frequent, though at the fame time they are preceded by little or no pain. With thefe fymptoms is fometimes (but not always) joined a tenefmus: the excrements are rather liquid than otherwife, fometimes without blood, and fometimes variegated with bloody ftreaks, though at times they are fomewhat hard, and encircled round about with blood and mucus. The appetite is, in the firft period of this dyfentery, twice or thrice greater than in health; however, it diminifhes by degrees, and at laft is quite deftroyed. The patient's ftrength does

not

not remain always at an equal pitch, but de-
creafes more and more by degrees, in the courfe
of the diftemper. In this manner it goes on for
three, four, fix, or even twelve months, (according to the obfervations made on it from the
year 1742 to 1748, by one Laurich, a German
phyfician, who has very well defcribed this diforder) and for the moft part ends in another
ficknefs, feldom in death. But we have alfo
nearer home, an example of an uncommonly
mild dyfentery, with which almoft all ftrangers
are feized during the firft week of their abode
in Paris. I have myfelf had this dyfentery fo
very like the Indian diforder, in Paris; it does
not even confine the patient in the leaft to his
bed, goes off in a few days, and does not even
deferve the name of dyfentery. We fee very
often in Switzerland, and in other particular
places, epidemic dyfenteries uncommonly benignant. In general, a dyfentery is called malignant, in which all the fymptoms that appear
at the beginning, are of much more confequence than they feem to be, or in which unufual fymptoms fupervene on a fudden, or
when the beft remedies, felected with the greateft deliberation, are without the leaft effect;
when many die without the leaft fault committed by the phyfician, the patient, or affiftants;
and when the fick, as Thucydides fays of the
Athenian

Athenian plague, die as well with, as without the phyfician.

Notwithftanding this, thefe fpecies of the dyfentery, partly on account of their frequent complication, and partly by reafon of their changeable and deceitful nature, are very often not nicely enough diftinguifhed in practice. Where the dyfentery rages very much, and kills a great many, there is ftill always a number of flight, and eafily curable cafes: and in malignant dyfenteric epidemies, there are likewife a great many fluxes in the fame place, that are not malignant. The characteriftic of malignity itfelf, differs vaftly in degree: in the city and diftrict of Zurich, reigned in the year 1746, a dyfentery, only in a fmall degree malignant. In the fame year raged in Saxony, a very malignant one, of which in a fmall fpace of ground and in a very fhort time, died an hundred perfons, moft of them on the third or fourth day, and none after the fourteenth. On the other hand, fymptoms of malignity may fupervene on a moderate dyfentery; it may even grow perfectly malignant, or elfe in many other ways dangerous. The benignant fpecies of dyfentery becomes contagious, malignant, and extremely dangerous, when many fick people are crouded together in a fmall fpace, or

where

where peculiar external or internal caufes, produce malignity in particular perfons. It is not
only poffible for a putrid fever to fupervene on
the flighteft dyfentery, but this fever may likewife terminate in a gangrene of various parts of
the body, at the fame time that the inteftines
are abfolutely free from any ailment of that
fort. However, moft dyfenteries end at laft in
a real mortification of the bowels, if they are
not properly treated directly at the beginning;
or when certain circumftances prevail, that render the beft method of cure abortive. But
when thefe do not take place, they go off very
eafily and terminate very happily, or elfe feem
uncommonly mild, merely on account of their
being taken care of in a proper manner. In
Sir John Pringle's opinion, a dyfentery that has
once got a good footing, is fo fixed and obftinate, that it can fcarce ever deferve to be called
benignant. But I fhall fhew hereafter, how
much the word malignant is abufed.

The dyfentery then is often more or lefs, either
of itfelf primarily malignant, or generally fpeaking, dangerous; merely according to the prefence or abfence of certain circumftances. It
appears in the army fometimes as early as the
fpring, when the troops have firft taken the
field; but is never fo obftinate, nor fo fre-

L   quent

quent as towards the end of fummer, or the beginning of autumn. At this time it grows epidemical and contagious, continues for about fix weeks or two months, and then ceafes; it is always more violent, if the troops lie wet in warm weather. It has alfo been obferved, that the dyfentery is always the more malignant, the earlier in the feafon it makes its appearance; and that it manifefts much lefs virulence, when it does not break out till Auguft or September. Otherwife I do not-fee that thefe camp-dyfenteries are in themfelves more malignant, than thofe that happen in cities; although in the army and military hofpitals, they become exceffively malignant and contagious from feveral circumftances; which, however, in like manner, take place in cities, when a great quantity of people attacked with this diforder, are crouded together in a fmall place, or where the other different caufes fubfift, of a peculiar or general malignity.

There are in all places dyfenteries, in which the diftinguifhing characteriftics of malignity are obfervable. A malignant dyfentery arifes, whenever that degeneration of the humours, which produces a malignant fever, is joined with thofe caufes, that generally occafion a dyfentery. Sometimes this conjunction proceeds from caufes, which are merely peculiar to one

or

or a few particular perfons, and then it only
gives rife to *fporadic* malignant dyfenteries.
Thus we generally fee, in epidemies of the malignant fever, here and there a fick perfon, who
is attacked with a malignant dyfentery; and in
epidemies of the mildeft dyfenteries, thofe
whofe juices are previoufly vitiated, are
feized with fluxes of the moft malignant kind,
or elfe fymptoms of malignity on various accounts mix themfelves with the more common
fpecies. The ordinary bilious dyfentery too may
become malignant, from violent or frequently
repeated fits of anger, as well as by means of a
bad method of cure; fuch as, for example, the
adminiftration of brandy and aftringents. Dr.
Tiffot faw a violent fever, that lafted fifteen
hours, enfue on a dyfenteric perfon's drinking a
large draught of brandy, at the fame time
his ftools entirely ceafed; after this febrile paroxyfm the patient loft all his ftrength, the dyfentery returned with an intolerable bad fmell,
he fainted away upon each ftool, his pulfe was
weak and very irregular, and he looked like a
perfect corpfe; he next went into a clammy
fweat, and died in forty-eight hours after drinking the brandy. I myfelf have feen and cured
a dyfentery, probably not naturally malignant,
but made fo by art, and have given a defcription of it in the former part of this work.
But when the conjunction of the general caufes

of

of the malignant fever, with the peculiar caufes
of the malignant dyfentery, has its origin in
circumftances of fuch an univerfal influence, as
to produce an epidemy, the certain confequence
of this is a perfectly epidemical malignant dyfentery ; that is, a great number of people are at
once, or very foon one after another, feized with
a malignant dyfentery. An epidemy of this fort,
is, next the plague, of all others the moft dangerous, and has likewife been feen in conjunction with the plague: unufual and extremely ftinking vapours, an extraordinary heat,
famine, or a camp pitched in marfhy grounds,
have often made this dreadful diftemper epidemic. Bontius faw it in Batavia, when this city
was befieged in the years 1724 and 1728, by
the people of Japan. In Europe it has been
obferved from 1548 to 1746, and ftill later,
and that principally in France, England, Germany, and Switzerland. In the year 1673, a fqua-
dron of horfe, confifting of fix hundred men, under the command of the marquis of Laffingen
in Flanders, that had lain too long in fwampy
grounds, was attacked with this dreadful diforder,
which was likewife attended with a mortification of the thighs ; five hundred and forty dra-
goons died of it, and at the fame time a great
number of horfes. Befides thefe manifeft caufes,
fuch an epidemy is often alfo an effect of thofe
degenerations of the air, which do not fall un-

der the cognizance of our fenfes, the terrible influence of which, however, we do but too fenfibly
feel.

Thefe remarks on the difference between thefe
fpecies of the bloody flux, amount to this ; that
without doubt there are benignant, and as inconteftably alfo, malignant dyfenteries ; but
that we fhould not fo readily look upon a dyfentery as benignant, when the fymptoms are not
bad directly at the beginning; becaufe in that
very cafe, influenced by certain circumftances,
every thing may turn out bad ; that it would,
therefore, be better not immediately to eftablifh for a certainty at the bed-fide of the patient,
what is not fo certainly eftablifhed by nature
herfelf.

It is juft the fame with its contagious nature.
The fame dyfentery is infecting, or not, according as it varies in circumftances. Without poffef-
fing of itfelf any real malignity, ftill a dyfentery often becomes truly peftilential in foul
and crouded hofpitals, and, of confequence, fo
much the more infectious. It is on this account,
that the dyfentery is fo common and mortal in
the field, and therefore, a conftant and vaftiy
dangerous camp-diftemper. The great havock
made by the dyfentery, proceeds always in the
army from the infection, that arifes from the pu-

trid

trid excrements of many people lying fick of it
at the fame time ; which, without proper meafures being taken, is likewife the cafe in the country villages, and fometimes even cities. As
mild too as this diforder may appear in the firft
days, yet with moft of the patients, before it
terminates in death, the excrements acquire a
cadaverous fmell, and are then exceffively contagious. I have found this fmell fo offenfive in
a woman of eighty-one years of age, during the
epidemy of 1766 at Brugg, that the continual
keeping open of the windows and door, and perpetual fcenting the room with vinegar, were not
fufficient to prevent its infection, and two of her
nurfes were feized with it. Now, fince a dyfentery
that proves mortal, is in a certain meafure always
of a contagious nature, though, to be fure, it does
not always communicate that contagion ; it follows hence, that every dyfentery is infectious, which reigns for any time in a country,
attacks a great number at once, and has proved
mortal to many. This is the more evident,
when one confiders how much fear affifts contagion in every refpect. In the malignant epidemic dyfentery of 1746, in the city of Zurich, it
generally happened, that many that were inhabitants of one and the fame houfe, perhaps chiefly
on this account, were feized within a few days
with this diforder, as foon as one perfon in the

houfe

houſe had got it ; for the very ſame reaſon all
ſuch as lived together in one houſe, grew much
worſe, and very often a great many of them
died, when any one in that houſe had previouſly given up the ghoſt. When hoſpitals are
filled with dyſenteric people, ſome of the aſſiſtants are attacked only with the dyſentery, and
others with the jail, or hoſpital-fever, that ends
in bloody and gangrenous ſtools ; in general
. likewiſe, almoſt all malignant fevers without exception infect the aſſiſtants, when there is not
proper care taken with regard to keeping the air
ſweet, and particularly in removing the putrid
ſtools. In the army, the long ſtay of the troops
in one place keeps up the havock of the dyſentery amongſt them, which ſometimes on the removal
of the camp, goes off on a ſudden ; nothing is
therefore more wholeſome for armies in ſuch
circumſtances, than plenty of exerciſe, and being at a diſtance from the houſe of office, the
foul ſtraw, and other naſtineſs of the camp.
From all theſe obſervations, made partly by me,
and partly by other phyſicians, I conclude, that
the dyſentery is very often only accidentally contagious, but that it alſo frequently becomes eſſentially ſo, juſt before the death of the
patient ; and that, in general, in all epidemic
dyſenteries, this diſorder, without the uſe of
proper preſervatives, muſt neceſſarily likewiſe be

ſpread

ſpread by contagion. But I cannot by any means admit with Degner, that contagion is the chief occaſional cauſe of this diſorder with every body.; although, without doubt, the in-ſection is eaſily communicated to linnen, beds and clothes, and thus produces the diſeaſe in ſuch, as are in a condition fit to receive it.

After theſe obſervations on the different genera of the dyſentery, I proceed to its ſpecies.. Juſt in the ſame manner as the former are com-plicated one with another, ſo are very often the latter: a dyſentery, attended with a putrid fe-ver, may either openly, or elſe in an undermin-ing and almoſt imperceptible manner, be ac-companied with inflammation; or the putrid fe-ver may be changed into a perfectly malignant one : and a dyſentery, attended with an inflam-matory, putrid, or even mild malignant fever, may, after all, turn out a chronic caſe. How-ever, this ſhould not hinder us from dividing the dyſentery into its proper ſpecies, ſince it ap-pears under ſo many different ſhapes, and ſhould be, without doubt, variouſly diſtributed, ac-cording to them, as under each form it muſt be variouſly treated. From this, a perſon of any penetration may perceive, the dreadful difficul-ties that attend the exerciſe of the medical art : as its objects, namely diſorders, are of ſo un-

ſtable

ftable a nature, fo liable to be miftaken one for
another, and even fometimes fo variable in their
fpecies.

In dividing the dyfentery into its divers fpecies, phyficians have been at all times extremely
liberal ; they have committed the fame fault, as
Hippocrates blamed in the people of Cnidos,
and in which M. de Sauvages is entirely involved
in his Nofologia, having made particular and
almoft fingular cafes of fo much importance, as
to defcribe them as fpecies. Degner is, in my
opinion, one of thofe, that have written the beft
on this diforder : I look on him as a very good
obferver, and, indeed, a very refpectable phyfician, but yet, not as a man of true genius ;
for it appears to me, that he was not very
capable of joining phænomena together in a proper manner, developing fufficiently complex
ideas, and ranging the heads in a natural
order. On the one hand, he has not well diftinguifhed our bilious, or, as it is called, putrid
dyfentery, from the malignant, though they
differ from each other, in his hiftory of the epidemy at Nimeguen, which was compounded of
both fpecies ; and on the other hand, fuppofes
the bloody flux, the dyfentery without blood,
and the mucous dyfentery, to be quite diftinct,
and different from the bilious. Other phyficians,

cians, even in this enlightened age, speak of a
grey dysentery ; of a dry dysentery, which I am
very well acquainted with, but have made no
particular species of it, and in which they very
justly advise lenitives and emollients ; and,
lastly, of an acid dysentery, which seldom happens, and principally affects weakly people :
but there is as little reason for making a specific
distinction between a bloody flux and a dysentery
without blood, as between a grey, yellow,
green, or black dysentery. Blood being mixed
with the excrements, is indeed an usual, but not
inseparable symptom of the dysentery ; for many
have all the other signs without this, at least in
the beginning, and others have blood in their
stools for various reasons, without having the
dysentery. But although this disorder is very
often attended with blood, it does not, however, on this account, deserve the name of the
bloody flux, as the appearance of blood here is not
an essential and inseparable token of a peculiar species : one may then have a genuine dysentery,
without any blood appearing in the stools ; and
a dysentery may be extremely dangerous, without any appearance of blood. The excrements
in a dysentery are likewise often only white, but
frequent experience has convinced me, that they
seldom remain of that colour during the whole
course of the disease ; and that these perfectly
white

white excrements in a genuine dyfenteric cafe, are not accompanied by any one fymptom, which may ferve to diftinguifh it from a dyfentery attended by a putrid fever. Formerly the dyfentery without blood, or white dyfentery, as it was called, was looked upon as much more dangerous than what is termed a bloody flux, or red dyfentery; as to the former, they afcribed a malignant nature, and confidered the excrements in this fpecies, rather as purulent, than mucous or watery; but I have fhewn in the fecond chapter of this work, that thefe purulent excrements are very often a mere chimera, and fhall alfo fhew in this chapter, that the appearance of danger in the dyfentery is taken from quite different tokens. The moft important and dreadful dyfenteries were not, even in thofe times, looked on as dyfenteries, if they were neither of the red nor white fpecies; for the phyficians, who publifhed the Breflaw obfervations, have put it as a matter of doubt, whether the painful diarrhœas, (as they call them) defcribed by Willis and La Moniere, entirely without blood, fhould be looked upon as real dyfenteries. This their doubt, feems to me to deferve compaffion. A great man in our art, Dr. Morgagni, teaches us, that whether a watery matter, howfoever coloured, or even . pure mucous, come away from the patient,

a great

a great many phyſicians, after Willis's and Sydenham's example, have neverthelefs, for a
long time back, always called ſuch affection a
dyſentery, if the ſtools were copious and very
painful, though, at the ſame time, entirely
without blood. This opinion, maintained by
Dr. Morgagni is, in my judgment confirmed by
the following conſiderations : firſt, as the dyſentery here in queſtion, obſerved by Willis in the year
1670 at London, brought the patients down fo
much within twelve hours, that they really ſeemed
to be on the brink of death, and, indeed, actually
died, if the phyſician did not haſte to affiſt them
with corroborant remedies, and omit evacuations
of all ſorts ; and ſecondly, as ſuch a dyſentery has
been obſerved to prove mortal on the thirteenth
day, while neither blood nor pus came away from
the patient, and even in the dead ſubject the inteſtines were found perfectly ſound. But an argument againſt theſe good gentlemen of Breſlaw,
that comes nearer to the point, and thruſts ſtill
more home, is this, that the dyſenteries de-
ſcribed by Willis and La Moniere, evidently
belong to the malignant ſpecies. All this,
taken together, I give it as my opinion, that
the various ſpecies of dyſentery are not to be
diſtinguiſhed by the difference of the evacuated
matter, but of the fever, by which they are accompanied.

But

But great phyſicians have likewiſe ſhewn, that there are many dyſenteries that do not come within the compaſs of a profeſſed treatiſe on the dyſentery, namely, ſuch as are only ſymptoms of another very different diſorder. After an inflammation of the ſtomach or bowels, ulcers, or even cancers, may be produced in them, the effect of which is a dyſentery of this kind. An abceſs in the liver yields a thin pus, mixed with blood and gall, that comes through the ductus choledochus into the inteſtines, and produces a ſpecies of dyſentery; acting in the ſame manner, as an ulcer in the pancreas, which pours through the pancreatic duct a purulent bloody matter into the inteſtines. A ſuppuration has been obſerved to ſpread (by means of blood-veſſels common to both) to the inteſtines from the meſentery, on which they are ſuſpended, when this latter has been inflamed and ſuppurated, or elſe it has affected them by metaſtaſis, and in both ways excited a very dangerous dyſentery. That painful kind of flux, called the hæmorrhoidal, which proceeds from the inferior part of the colon, or the ſuperior of the rectum, is often taken by ignorant people for a real dyſentery, as it has ſome reſemblance with it. All ſorts of corroding and irritating acrid matter, whether taken in at the mouth, as having its origin in the body itſelf, from corrupted hu-

mours

mours flowing into the inteftines, produces a
fpecies of dyfentery. Bloody fluxes arife, as
well as other hæmorrhages, after the amputa-
tion of any member. Sometimes a dyfentery is
a fymptom of an intermittent fever. There are
petechial fevers, in which a genuine dyfentery is
likewife a fymptom at the beginning. It alfo very
often terminates, or is a concomitant fymptom
of putrid and malignant fevers; but when a malignant fever fupervenes on a dyfentery before
fubfifting in the body, this is quite a different
cafe, and conftitutes a peculiar fpecies of dyfentery. In military hofpitals the dyfentery is
complicated alfo with other diforders, particularly with coughs and pneumonic affections,
when the weather fets in for cold. In the fcurvy
it is a very dangerous fymptom.

However, we muft diftinguifh from all thefe
fymptomatical fpecies fuch, as do not depend
upon another diforder, and of thefe only we here
treat at prefent, and fhall take four fpecies into
confideration, though there are, perhaps,
more; if there be, however, they are but feldom feen. The moft ufual are, the fpecies
which accompanies an inflammatory fever; that
which attends a bilious or putrid fever, the moft
common of any; that which is concomitant on

5   a malig-

a malignant fever; and, laftly, (if it may be admitted) the chronic dyfentery.

The dyfentery appears fometimes with a re-gular inflammatory fever, a hard and full pulfe, a very violent head-ach, and a diftended belly. An Englifh phyfician, Dr. Aken-fide, feems indeed not to believe the exiftence of that fpecies of dyfentery, which is attended with an inflammatory fever, affirming, that all fevers of this kind terminate in an inflammation of the bowels, which does not produce any diarrhœa in the leaft. He adds alfo, that there are no other ulcers in the dyfentery, than thofe that are the effects, and not the caufes, of the malady; and on the ftrength of this, he thinks he has a right to fall foul on Boerhaave, for not having attended fufficiently at the bed-fide of the fick; but inftead of that, confidently explained the caufes of things from the chair to his difciples. Dr. Akenfide is in the right in confidering the ulcers of the inteftines, that are remarked in the dyfentery, merely as an effect of the dif-temper; for if they were the caufes of it, it would then belong to one of the fymptomatic fpecies before defcribed. Dr. Akenfide would have been in the right if he had faid, that a common inflammation, or even ulcer, in any part of the bowels, does not produce a dyfen-tery;

tery; but when he denies, that an inflammatory ftate of the blood, which has almoft the fame effect on the bowels, as an inflammation of the eyes has on the tunica albuginea, may produce an inflammatory dyfentery, he denies a thing which is demonftrated, though feldom feen in London; and argues directly, as if one fhould fay, this man fpits blood, and confequently has no inflammation in the thorax. Dr. Aken-fide is alfo in the wrong, in committing the very fame fault himfelf, with which he fo unjuftly reproaches Boerhaave; for it is very certain that we may fee, in the courfe of practice, real dyfenteries of an inflammatory fpecies, which Dr. Akenfide did not fee from his profefforial chair, and in which there cannot poffibly be any conftipation of the bowels, as the ftools in thofe cafes are at times uncommonly frequent, though accompanied with very little excrement. In September 1734, this fpecies of dyfentery fhewed itfelf in the village of Viterne in Lorrain; it came on with a breaking of wind up-wards, and a very violent pain in the ftomach and bowels; on this enfued a fever, and foon after frequent dyfenteric ftools, with a tenefmus and unquenchable thirft, and fuch an inflamma-tion from the œfophagus downwards to the anus, that the patients thought their infides were on fire; the tongue near the œfophagus was

inflamed

inflamed and black : if the patient vomited in this condition, he died on a fudden. In the fpace of ten days, fifteen perfons died of this diftemper, defcribed by Dr. Marquet, dean of the college of phyficians at Nancy. Some who were feen walking about the ftreets at five o'clock in the afternoon, were feized with it, and died at ten o'clock at night.

I fay nothing in this place of that fpecies, which is attended with a putrid fever ; as this is the very fame fpecies, on the fubject of which we have given fome hints in the firft chapter of this work ; for, ftrictly fpeaking, it would be improper to fay that they are therein de-fcribed. Perhaps many additions, taken from my later obfervations in the year 1766, and difperfed up and down in this chapter, may fill up part of thefe chafms, and are fufficient for the end I propofed, as the whole treatife itfelf is nothing but patch-work.

We do not give the title of malignant to, nor make a peculiar fpecies of every dyfenteric cafe, in which fymptoms of an alarming nature break out on a fudden, where the beft remedies, chofen with the greateft judgment, have no effect, in which many of the fick die, and a great number of thefe very faft, and almoft on a fudden ; and

M   where

where they die, as well with, as without the affiftance of the phyfician. A dyfentery of this
kind may, notwithftanding all thefe circumftances,
be of the inflammatory fort : it is, to be fure, extremely violent, dangerous, and alarming ; but,
ftrictly fpeaking, it ought not, even then, to be
called malignant, as it is important to employ this
word in a quite different fenfe. A philofophical
phyfician entertains the more ftrict and determinate notion of malignity with refpect to the
dyfentery, in cafe only, that with the caufes
common to that diforder at all times, and in all
countries, others ftill are joined, which corrupt
the humours very quickly : it is this circumftance, that impreffes on a dyfenteric complaint
the peculiar mark of malignity, and then conftitutes that peculiar fpecies, which we are now
about to treat of. The malignant dyfentery is,
therefore that, in which either from external
caufes, or from a putrid fomes within the
body, a malignant fever fupervenes. The
pathognomonic figns of this fpecies are formed
by the fymptoms of a malignant fever mixed
with the ufual fymptoms of the dyfentery with
different degrees of force, and are chiefly
alarming on account of their complication.

The moft important fymptoms of the malignant dyfentery are, befides the ufual fhiverings,
(which,

(which, though not always prefent, often return
in the courfe of the diforder) a fudden total lofs
of ftrength, and an extraordinary anxiety about
the region of the pit of the ftomach ; this remains till the end of the diforder, when it proves
mortal, or till a remarkable change for the better, when the patient is deftined to recover : it
does not let him fleep a moment, though otherwife he fhews a good deal of infenfibility with refpect to every thing, and even to his own
diforder ; for the moft part his head is heavy
and ftupefied, but at times fo intolerably painful, that his fcull feems perfectly fplit afunder ;
he is often at the beginning of the difeafe in a
calm delirium, which manifefts itfelf particularly
by an extraordinarily wild glance, a fixed ex-
ftatic look, appearing in very deep thought,
while, at the fame time, he thinks on nothing
at all. At times this delirium grows very high,
the voice moft commonly changes, and becomes
weak ; a flight impediment in the fwallow is
often obferved juft at the beginning of the
diforder, which is an extremely bad fign. It
happens very commonly, that the patient vomits
worms, or that they are found in his excrements, or elfe that they creep up of themfelves
into his palate and mouth, and fometimes even
into his noftrils, fo that he can pull them out
with his fingers; but we muft take care not to

M 2   confider

confider worms as a fign of the malignant dyfentery, as they are feen in great quantity alfo in
many epidemies of the bilious dyfentery : a copious vomiting of a perfectly green matter often
appears, without any relief following from it;
fometimes too, the patient vomits blood, which
is a very bad fymptom.   The pains in the
bowels are not always proportioned to the danger of the diforder : there are fome who do not
complain in the leaft of them ; with others they
are extremely violent, and with others intolerable.   Sometimes the belly remains foft, and
fometimes it is diftended ; the ftools are fometimes inconceivably copious, and this is fo very
dangerous a circumftance, that the patient will
appear in a dying condition in the fpace of twelve
hours, and often really dies in that time.   Sometimes hardly any thing at all comes away from
him ; he has not the leaft complaint, except a
very troublefome tenefmus, and dies in three or
four days, if his ftools cannot be rendered
more liquid.   The excrements are fubject to
great variations : fometimes perfectly mucous,
fometimes of a dark brown, citron-yellow, or
green colour; at times they confift of mere
water, and are at the fame time amazingly copious; fometimes entirely of thin watery blood.
At this juncture, the patient is obferved to grow
worfe and worfe every hour, is light-headed,

feems

feems to fuffer very little, and dies on the third day. Sometimes the excrements are of a greyifh red; they are often black, though ftill more commonly flimy, mixed with a fubftance that looks like chocolate and blood, and has always an extremely bad fmell. Copious ftools, with a low finking pulfe and increafing delirium, are bad figns ; on the other hand, it is a good token to have bilious ftools followed by a fweat: in flighter cafes, a fweat often puts an end to the alvine flux, and the other fymptoms almoft on a fudden ; while, on the contrary, the entire abfence of this fweat is dangerous. Heat of urine, and the ftrangury, are ftill more frequent than in the common bilious dyfentery : as thefe fymptoms are in general obferved di-• rectly at the beginning of malignant fevers, they are a bad omen in the malignant dyfentery, in which there is alfo fometimes an entire ftoppage of the urine. This fluid is fometimes perfectly brown, a circumftance which indicates death ; fometimes it is as clear as water, and fometimes milky : the bad fmell of the urine approaches fometimes to that of the excrements. The fame has been alfo obferved in the breath, in the fputa, and even in the fweat. The patient's difguft for all kinds of food is unconquerable, and often, with the greateft thirft upon him, he cannot abide any drink, that is not cordial : fometimes

M 3   the

the fkin is extremely dry, and peels away in
great pieces, or elfe it is conftantly cold and
clammy.   In an epidemic diforder of this fort in
France, it was remarked, that fuch as had a
great number of watery bladders break out on
the whole furface of the fkin, recovered : the
fame good effects were obferved in Switzerland,
when a miliary eruption appeared, and continued till the feventh day, if, at the fame time,
tumors appeared here and there on the fkin,
together with an eryfipelas.   In other epidemies, juft before the approach of death, petechiæ and large miliary puftules have been obferved to break out.   The petechiæ fhew themfelves very often on the fourth, fifth, fixth, or
feventh days, though they are not conftantly
connected with this kind of dyfentery : they appear moftly on the breaft, back, arms and legs ;
rarely, and almoft never in the face.   They
are not to be reckoned in the number of mortal
fymptoms, but they help, with other fymptoms, to increafe the danger ; and the deeper
colour they are of, fo much the worfe they are.
I have feen them, in a cafe that proved mortal,
of a brown, and even a bluifh hue, in an inconceivable number all over the body : fpots
and puftules have been likewife feen on the neck,
under the arm-pits, loins, and about the groin,
that were turgid with a green pus, and manifefted

fefted a peftilential nature. The only characteriftic of the pulfe is, fmallnefs; and it is very
feldom that the refpiration is not oppreffed. Immediately at the beginning, a hiccough, an increafing difficulty in fwallowing, a diftenfion of
the belly, a perfect drynefs and blacknefs of the
tongue, faintings, fometimes gangrenous fpots
on various parts of the body, and particularly
on the legs and feet, denounce, for the moft
part, a near and inevitable death. However,
in fome inftances, even a mortification, that has
towards the end of the diforder fuddenly feized
on the feet, has been found curable; and in
others, common inflammatory tumors on the
arms, as well as erifipelatous eruptions, that
came to a fuppuration on the legs, have been
obferved to be critical, and eafily healed; though,
at the fame time, the fkin was covered with miliary and petechial exanthemata: the ceffation
of the patient's anxiety, the foftnefs of his belly,
the free paffage of his urine, the diminution of his'
weaknefs, and particularly the return of his natural fleep, give certain affurances of a happy
cure, which is often effected, to the great honour
and triumph of our art; but alfo, full as often,
cannot be brought about by the beft phyficians.

The flow, or chronic dyfentery, does not
conftitute a peculiar fpecies, till the diforder has

made

made fome progrefs ; for though one may even in the beginning, from fome peculiar tokens, forefee its long duration, it is not yet even then the queftion, whether it fhould be treated in the fame manner as a chronic dyfentery : for fuch a queftion would be rather too fimple.

We call a dyfentery flow or chronic ; when three or four weeks are paffed fince the firft attack, without our having any hopes of feeing the diforder cured : it lafts very often many months, and fometimes whole years ; for at the very time, that I am writing this, an old man is come to me, that has had the dyfentery already two years together, with the ufual evacuations, and, notwithftanding that, ftill goes about, and does fome work. This malady arifes from an utter negleƈt of all remedies in the beginning of the diftemper, as well as from leaving them off too foon ; great errors in diet, a bad method of cure and frequent relapfes, and fometimes becaufe, on account of the preceding bad condition of the bowels, the beft remedies are without effeƈt. In this fpecies, the patient is very weary, his appetite very fmall, and his indigeftion fo great, that not only a violent pain in the ftomach enfues upon eating moft kinds of food, but alfo his food comes away perfeƈtly undigefted, as in the lientery. The pulfe is very weak and

flow ;

flow; but when there chance to be abfceffes or
fuppurations in any part, it is quick. The
ftools are, to be fure, not fo copious as in the
beginning of the illnefs, neither are they quite
fo painful, nor attended with fuch frequent attacks of the belly-ach. The excrements are, in
general, juft as in the firft period of the difeafe,
fometimes accompanied with blood, and fometimes entirely free from it: nay, there is frequently, even for fome years after, an appearance
of blood in them, or even of genuine pus,
when abfceffes in the ftomach or in the inteftines chance to break, or when the patient has
obftinate ex-ulcerations in the bowels ; and when
there is an ulcer of this kind in thofe parts, the
excrements confift of a thin, fharp, foetid, and
cancerous matter. All thefe chronic cafes of the
dyfentery are extremely obftinate, and with
many people mortal : they likewife terminate in
other diforders, and chiefly in the dropfy ; and
are never cured without extraordinary patience,
obedience, and refolution on the part of the patient.

After this fhort view of the moft ufual fpecies
of the dyfentery, it ftill appears worth while to
collect the moft common fymptoms (for, to be
fure, in this compendium, a great part of them
muft neceffarily be omitted) belonging to
all

all thefe different fpecies, with their various changes
and terminations ; (though I do not intend to repeat what I have already in part treated of, efpecially with regard to the malignant dyfentery.

An inflammatory dyfentery makes its appearance, at the beginning, with a very violent fever, a very hard pulfe, which, in other dyfenteries, is moftly fmall, and but feldom (and
that only in the progrefs of the ficknefs) becomes full ; an almoft continual and intolerable
pain in the belly, which increafes on the part's
being touched, and ftill more after vomiting ;
ftools very inconfiderable with refpect to quantity, a head-ach, red face, and fometimes a diftended abdomen.   A putrid dyfentery difcovers
itfelf by a bitternefs in the mouth, that appears
directly on the firft attack ; a vomiting of a bilious matter, which is fometimes alfo mingled
with worms, a fhivering that returns in the
courfe of the diforder, the fometimes appa-
rent flightnefs of the fever, the commonly pale
colour of the countenance, the eafe that is found
after vomiting, the variegated colour of the excrements, and fometimes by the worms con-
tained in them.   We may always fuppofe à
priori the prefence of a malignant dyfentery ;
where many people, fick of the dyfentery, are
crouded together in a fmall fpace ; but this
diforder

diforder may likewife proceed from many other external, as well as internal, caufes : its fureft pathognomonic figns are, the quick approach of a more than natural weaknefs, great anxiety about the pit of the ftomach, a heavinefs in the head, a wild, and yet at the fame time, a dead-like look, fpirits extremely depreffed, or a perfect indifference to every thing in the world, fre- quent flight convulfions, a very weak voice, a great many fainting fits, fometimes a miliary eruption, petechiæ, aphthæ, a very weak pulfe, a vaft ficknefs at the ftomach, and the other ufual fymptoms in malignant fevers, which have been above related. The flow, or chronic dyfentery, is manifeft enough of itfelf, and re- quires no defcription.

A vomiting of a copious bilious matter, that comes on of itfelf in the beginning, is of fervice in the putrid dyfentery ; but one that returns very often in the courfe of the diforder, and al- ways enfues, whenever the patient takes any thing, even in the fmalleft quantity, is quite the contrary : the leaft vomiting, even in the be- ginning of the diforder, in an inflammatory dy- fentery, is bad ; a hiccough in the beginning is of very little confequence, when it proceeds from indigeftion, wind, or worms : but when the diftemper is at its height, or has continued

any

any time, it is a token of an inflammation,
and an approaching gangrene. It is a good
fign, when the patient fhews an appetite; the
want of it, with an increafing difguft to food,
is bad.   Frequent ftools, with a fmall evacuation, is the ufual complaint; however, the
diforder is always worfe, the more frequent
and fmaller the ftools are, and the more frequent and fruitlefs the tenefmus is in the firft
days of the illnefs.   Large and few ftools are
good; large, very frequent, and fuch as do
not in the leaft diminifh the diforder, are bad,
and a token of a very powerful irritation in
the bowels.   Large ftools, when the diforder
has continued any time, are bad, if, at the
fame time, the food comes away undigefted, or
if, without this circumftance happening, they
are only frequent.   Streaks of blood fhew a di-
laceration of fome fmall veffels in the rectum,
which is of no confequence; a great excretion
of blood proceeds moftly from the fame part,
or from the lower end of the colon, the evacuation of which is likewife, in other cafes, fo.
very innocent.   I faw, very lately, a great deal
of blood come away in fome dyfenteric complaints, that yet were otherwife, notwithftanding the fever and the frequency of the ftools,
very tolerable, and almoft entirely without
pain.   Hence it is, that fo many obfervers have
                              found

found a large excretion of pure blood to be of
no difadvantage in the dyfentery, but rather of
fervice; while, on the other hand, many of
their patients died in a fhort time without theleaft
indication of blood in their excrements. Thofe
who loft blood in great quantity, were, according to Degner's obfervations, always in lefs
danger, than thofe that had but little pure
blood in their excrements, but, inftead of that,
a white, frothy, tough mucus only ftreaked
with blood; as thefe laft complained of more
violent pains, more frequent ftools, and a
greater lofs of ftrength. A more intimate
mixture of blood with the excrements, is
looked on as a fign, that the blood comes from
a higher part than the rectum, and phyficians
are, upon that account, very much afraid of
this token : without doubt, we may hence conclude, that the diforder exerts itfelf chiefly in
the inteftina tenuia, where there mu⬛certainly
be greater danger. In reality, I have ⬛this
thorough mixture of blood with the excrements, in patients that were in extreme danger;
but I have likewife feen it in very flight, and
very eafily curable cafes. Towards the approach of death, the ftools became lefs bloody,
that is, lefs red; for the blood is then changed
into a putrid, ichorous matter : in general, the
danger in this diforder is not in proportion to
the

the quantity of the evacuated blood ; and it is
only in the malignant fpecies, that any lofs of
blood whatfoever is, for the moft part, extremely dangerous. With regard to the other
figns taken from the excrements, there are very
great errors committed, not fo much on account
of the pus being taken for a mucus, as by too li-
berally beftowing the name of pus on mere mucus : in general, the more the colour of the
ftools departs from that, which is natural to
them, the worfe they are ; a green colour is a
fign of a perfectly vitiated gall, black is the
worft of all. The excrements have always a
putrid fmell ; but when a gangrene fupervenes,
they prove perfectly cadaverous, though this
may happen alfo juft before, and at that time
they are more infectious than at any other.
Worms, which I faw in great plenty in the epidemy of 1766, in children as well as adults,
make the dyfentery worfe, in the fame manner
as they do the putrid fever, and are fometimes
found in the excrements, and, at other times,
brought away by vomits. They are, for the
moft part, of the round kind, though I have
likewife feen in the epidemy of 1766, a perfectly
aftonifhing quantity of afcarides ; but, according to Sir John Pringle's important admonitions, worms, in general, fhould by no means
be confidered as the caufe of this diforder, but

merely

merely as a token of the preceding bad condition
of the inteftines, of their tone being weakened,
of a diminution of the natural fecretions, and
of a coagulation and corruption of the aliment.
Aphthæ often come in the mouth and on the
tongue fuddenly, and are very dangerous, as
well as a difficulty in fwallowing; the belly-
ach is always more dangerous, the more violent, and longer continued it is, and the lefs it
ceafes after evacuation.   An ardent heat in the
abdomen, a fcalding in the urine, and even
the ftrangury itfelf, are in the bilious dyfentery
only a fign of an irritation from the gall, and
are attended with no bad confequence; in the
malignant dyfentery, thefe fymptoms are a-
mong the dangerous ones.   An anxiety about
the breaft and pit of the ftomach, are obferved
in divers fpecies of this diforder, and is always
very dangerous : all hyfterical fymptoms muft
be treated as fuch, and therefore fhould not be
abfolutely confidered as immediately proceeding
from the diforder; but all thefe fymptoms are
bad in a malignant dyfentery : real convulfions
in a common dyfentery, after the ufe of aftringents, are fatal with children.   All nervous
fymptoms are dangerous in every fpecies of this
diforder; as they are always a fign of a very
powerful irritation in the bowels, and fome-
time kill them almoft on a fudden.   Miliary,

or

or petechial eruptions, are moftly dangerous ;
and though they are not reckoned among the
abfolutely mortal fymptoms, yet they help very
much to increafe the danger.

The diftemper grows in general very dangerous, when either through neglect, or bad
conduct, it lafts till the patient's ftrength is
quite gone, the inteftines relaxed, and the tunica villofa abraded ; though there are ftill
hopes, as long as there are neither ftools confifting of diluted blood, or coming away invo-
luntarily, nor aphthæ, nor petechiæ, nor a hiccough, nor any complaints of a great weaknefs
and anxiety about the præcordia ; in which
cafes, the beft practitioners lofe all hopes. A
complication of many dangerous tokens, is a
fure mark of approaching death, though many
of them confidered in themfelves, and fingly,
by no means denounce death ; fuch are faintings, hiccough, vomitings, particularly of an
unufual matter, heart-burn, anxiety about the
præcordia, green ftools, afh-colour ftools, or
confifting of a dilute, watery blood, and not
tinged with real blood ; ftools mixed with
worms, and of an extremely cadaverous fmell ;
a conftant rejection of clyfters as foon as taken,
extraordinary wakefulnefs, unquenchable thirft,
cold at the extremities, a perfect lofs of ftrength,
a weak

a weak finking pulfe, a flight fever of the malignant kind, or elfe, to all appearance, an entire abfence of fever, an internal burning heat,
cold fweats, aphthæ in the mouth, which grow
black, a difficulty in fwallowing, a gangrenous
fore throat, perfectly brown fputa, a fudden
ceffation of all pain, retention of the urine, a
defire of getting up, a great inclination for cold
water, livid lips, heavinefs in the head, a flight
delirium, fubfultus tendinum, convulfive 'motions of the whole body, the eyes funk in the
head, a wild look, and involuntary ftools.
The event is always very dubious, when the
violence of the diforder has not been allayed by
proper evacuations, as a gangrene very often
flowly follows fuch neglect; on the other hand,
the happy termination of the diforder depends on
the fpeedy ufe of the medicines appropriated to
each fpecies, and applied at a time when the
patient's ftrength is not as yet too much exhaufted, nor the inteftines in the leaft damaged.
No illnefs is more fubject to relapfe than the
dyfentery, and frequent relapfes occafion a continual diarrhœa, as they weaken the tone of the
inteftines, erode their tunica villofa, and even
occafion ulcers in them. The figns of the diftemper's having a happy exit, are the ceffation of all the fymptoms, that appeared at the
beginning, with many other phænomena which,

N   a man

a man with a very fmall fhare of judgment may
eafily perceive.

Thefe various circumftances require various
methods of cure. There are fpecies of the dyfentery, in which remedies, that are found to be
ferviceable in a different fpecies, prove fatal; and
in which, on the other hand, thofe that prove
mortal in another fpecies, are of the greateft
fervice. There are even dyfenteries of the
fame fpecies, in which the fame method that
was ferviceable in cafes that feemed perfectly
fimilar, has a quite different and oppofite effect.
It muft, therefore, be very fenfelefs to employ
the fame remedy in all cafes of the dyfentery:
it muft, therefore, be very fenfelefs to take it
into one's head, that a fpecific is to be had for
all fpecies of the dyfentery, or that fuch fpecifics are to be found in books, in which all the
fymptoms of the moft different fpecies are confounded together in the moft indigefted manner;
and it is, therefore, in like manner impoffible
to determine on an univerfal method of cure
for the various fpecies and periods of this
diforder. But alfo after the moft accurate dif-
tribution of its various fpecies, and their different methods of cure, much ftill remains here,
as well as in the whole circle of medicine; the
knowledge of which, though it cannot be tranf-

mitted

mitted by written precepts, is however of unfpeakable importance, namely, the manifold and
almoft innumerable complications of cafes,
which are only determined by the particular
circumftances of the patient. But, when once
informed of the nature and method of cure of
each fpecies, every phyfician muft be in a condition to advife himfelf in the moft complicated
cafes; if he be poffeffed of that genius, which is
beft tried at the bed-fide of the patient, and is
of itfelf every thing in medicine.

In the inflammatory dyfentery, venæfeftion in
the beginning is a principal article ; and there
is no occafion to fear repeating it, if the patient
has ftill ftrength, and is not too much exhaufted
by copious ftools : it has fometimes an aftonifhing quick and good effeft. After that fhould
be given, every day, three or four clyfters of
barley-water, marfh and common mallows,
and chamomile-flowers; but it is of great importance not to injeft the whole clyfter at once,
but at different injeftions, as it were, that it
may ftay the better, and not come away immediately without effeft: internally, foft, emollient, and mucilaginous remedies are of ufe ;
gum arabic, pulvis è tragacantha according
to the London Difpenfatory, fyrup of marfh-
mallows, and, with thefe, the frequent ufe of

warm

warm almond-emulfion, or barley-water. Warm fomentations of chamomile-flowers boiled in milk, are to be applied to the whole abdomen, after having rubbed it previoufly over and over with ointment of marfh-mallows. When the inflammation is perfectly cured, we may make ufe of the tincture of rhubarb with water, in fmall dofes, going on at the fame time with the almond-emulfion.

From a violent inflammation of the rectum, which very eafily comes to a fuppuration, the figns of which are obferved in the patient's ftools, an extremely troublefome tenefmus may arife. This fymptom is cured by venæfec-tion, and often repeated clyfters, but efpecially by leaches.

I have been of opinion, and probably moft phyficians in Europe, that all medicines, with-out any exception, are of no ufe; when an in-flammation of the inteftines is followed by a gangrene in thofe parts, and that fuch a gan-grene was always abfolutely mortal ; but I have fince found the juice of lobfters in broths and clyfters, recommended in Mr. Rahn's famous work on the dyfentery, in cafes, that had all the tokens of an internal gangrene.

Emetics

Emetics are, in this fpecies of the dyfentery, a deadly poifon. Purges, by their irritation, do not do lefs mifchief, as they increafe the inflammation. All opiates, heating, aftringent and binding medicines, are here extremely hurtful. It is not uncommon for phyficians of great reputation, after having found the true indication towards the cure of the diftemper, to chufe, for executing their purpofe, remedies that have a quite oppofite effect, or at leaft appear to have it. The Breflaw phyficians laid it down as an indication towards curing the dyfentery, that the inflammation fhould be refolved; with this view, they advifed radix tormentillæ, valerian, the confectio ex hyacintho, pulvis Hungaricus contra peftem, and even nutmegs; that is to fay, remedies that are aftringent, conftipating, heating, and confequently increafe the inflammation. Degner (with a view likewife to take away the flight inflammatory diathefis remaining towards the end of the dyfentery) even advifes the tincture of cafcarilla, which nobody will ever give as a good remedy, in a genuine inflammation of the inteftines. He cites alfo others to prove, that the radix pimpinellæ albæ is of fpecial ufe in the dyfentery, when the bowels are inflamed; though this root is acrid, hot, and irritating. Our excellent countryman, Mr. Rahn, in one part of his

N 3   work,

work, cautions his readers very earneftly againft the ufe of anodynes and aftringents, and in another place he advifes, (to be fure, for very weighty reafons) even when an inflammation is to be apprehended, Sydenham's laudanum, the fpecies ex hyacintho, and pillulæ ex cynogloffo, that is to fay, the very fame remedies which he cautions his readers againft. One great inconvenience in the adminiftration of opium is particularly this, that during the ufe of it, the inflammation gets to a head, without either the patient or phyfician knowing any thing of the matter.

The faccharum faturni is of fervice in external inflammations; Rivinus and Dolæus have therefore made trial of it in inflammatory dyfenteries, and thought it had a good effect. Now the faccharum faturni is, in general, a medicine, whofe bad qualities have rendered it juftly formidable, but whofe falutary properties (notwithftanding Mr. Goulard's experiments) are not yet fufficiently known, and it promifes not a little to phyficians of genius, who intend to make trial of it; as we may learn from the important obfervations of Dr. Hoze, a Swifs phyfician of extraordinary ingenuity, experience and merit. It cannot, however, be admitted in the dyfentery, as it confines in the
ftools

ftools and increafes the pain, and confequently
the inflammation.

Nothing can be more murderous, than the
method which Marquet advifed in the above-
defcribed inflammatory dyfentery, obferved at
Lorrain. He forbad venefeftion, and prefcribed ipecacuanha, rhubarb, diafcordium,
and a decoftion chiefly confifting of aftringents. If thefe counfels were of fervice, the
dean of Nancy's defcription of that epidemy is
fiftitious; if, like the plague, they occafioned
vaft ravages, and death, they did nothing but
what might be expefted from them.

In the bilious, or, as it is called, putrid dyfentery, the method may be purfued, which I
have defcribed in the firft chapters of this work
from my own experience, and towards con-
firming which, fomething more may here be
added ; but alfo a great deal more may be done,
or let alone, which I have not hitherto
been able to fpeak of, and muft, therefore,
make amends for it at prefent. To all this I
fhall add various obfervations made in the epidemy of 1766, and relate them always with
fincerity; though they fhould hurt my reputation, by fhewing when I have been unfortunate ; for a medical book is not worth taking

N 4   into

into one's hand, if the author does not fhew a
moft fuperftitious regard to truth.

Sydenham and Huxham have ordered venæfeċtion in general, at the beginning of the
diforder. Monro found it, in the Englifh army
in Germany, during the laft war, extremely
neceffary in recent cafes, and of very great ufe
in alleviating and curing the diforder : but
when the diftemper had lafted fome time, and
began already to grow tedious ; when the fever
was off, and the patient very weak, he looked
upon it as unneceffary, and even thought it did
harm.   Pringle,   indeed,   lays it down as a
maxim, that the dyfentery of itfelf does not
ftand in need of venæfeċtion ; but that when it
is accompanied either with a plethora or inflammatory fymptoms, he then looks upon
blood-letting as often indifpenfible, and always
of fervice towards the cure : yet, when the firft-
drawn blood has no tokens of an inflammatory
diathefis, and the fever is not fuftained by an extraordinary inflammation, he then finds it either unneceffary or hurtful to repeat it, as the
patient's ftrength muft be kept up in fo debili-
tating and putrid a diforder ; but all this is manifeftly meant in thofe cafes, in which an inflammation is joined to the putrid nature of
the diforder, and fo far I am perfectly of the
fame

same opinion with these great English
physicians.

Now, inflammation and putridity, or, (if
you will) a bilious degeneration of the fluids,
are not always found together : it is, therefore,
worth while to hear the Dutch and German
physicians on the excellence of their method.
Degner says, that in consequence of Sydenham's having called the dysentery a fever turned
inwards on the bowels, he advised blood-letting, in order to drive away from those parts
this afflux of acrid humours ; but if the dysentery must be called a fever, it naturally merits
the name of a sh——fever; as it expels, through
the posteriors, the whole substance of the body.
In the dysentery of Nimeguen, venæsection was
not necessary; accordingly, Degner did not advise any body to it; as it does not correct the
bile, but rather weakens the vis vitæ, and dis-
turbs nature in her salutary motions : accordingly he likewise saw venæsection followed
by vomiting of blood, and death. On these,
and other considerations, venæsection appears
to be very much suspected by Dr. Degner: as
nature does not easily bear two such evacuations at once ; and as taking all things
together, it is very little serviceable : he
found it rather more so in plethoric persons,

by

by way of prefervative ; and yet feems to think,
that in all cafes. one muft proceed in it with
great judgment and penetration, or, otherwife,
one runs the rifk of doing more harm than
good. Eller fays, we fhould in the beginning,
and the firft period of the dyfentery, examine
whether the patient be plethoric, and have a
quick and full pulfe : in this cafe, we may take
away fome blood, as it feems to be in a ftate of
inflammation, and even repeat this evacuation,
if the blood be covered with a yellow cruft,
which happens, however, extremely feldom;
but when there are no figns of a plethora, the
letting of blood is not only fuperfluous, but
alfo pernicious, as it takes away the patient's.
ftrength, fo neceffary to overcome the diftemper, of which he has, at the fame time, manifeftly very little to fpare.   I conclude then from
all this, that venæfection may, and fhould be
omitted in a dyfentery, that is merely attended
with a bilious fever alone ; but that there is
likewife not the leaft reafon to blame the phyficians, that make ufe of it in complicated
cafes.

Emetics, like purgatives, have been formerly either entirely neglected in the dyfentery,
or elfe too fparingly made ufe of ; and, neverthelefs, later experiments fhew, according to
Dr.

THE DYSENTERY. 187

Dr. Pringle's opinion, that they conftitute the
chief part of the cure. Eller fays, he has
found, (and it is confirmed by the long experience of phyficians) that no evacuation conduces
more towards the cure of an epidemic dyfentery,
than thofe that are cured by emetics. A
dreadful dyfentery raged in Auguft 1721, in
different parts of Upper Saxony. Eller was
called in it, and foon found, that no medicines
were fo effectual in the beginning of the diforder,
as thofe that at repeated times thoroughly expelled the irritating bilious matter; fo that by his
indefatigable pains he cured his patients moft
commonly within a fortnight or three weeks,
merely by evacuations, and even cured relapfes
in the fame manner. Out of three hundred
fick, which he had in his care, he loft fcarce
one man in forty.

It has been obferved, that emetics, with refpect to the evacuation of the bile, are more effectual, and that they have the beft fuccefs,,
when they work likewife by ftool. Both thefe
effects were obtained with more certainty by Dr.
Pringle in the Englifh army, when he gave,
inftead of the ufual quantity, only five grains of
ipecacuanha at once, and repeated it two or
three times the fame day, till a vomiting or
purging enfued, which generally happened either

ther before, or juſt after the third doſe. Fifteen grains, adminiſtered in this manner, procured a larger evacuation, than thirty grains given at once : however, though the Doctor found this manner of ordering the ipecacuanha very uſeful, eſpecially when it is repeated once, or oftener, after having been omitted a day or two; he left it off, on account of the extraordinary ſickneſs of ſtomach that followed upon it; though he imagined it to be the ſureſt method of cure. Dr. Eller made uſe of this method ſo long ago as the year 1721 ; ſo that he gave twice, thrice, or even four times a day, four, five, or ſix grains of ipecacuanha, till a gentle vomiting enſued : for ſtrong people, he mixed with each drachm of ipecacuanha four grains of emetic tartar, and of this compoſition he preſcribed four, five, or ſix grains, to be taken in the ſame manner at different times, with the beſt ſucceſs. Dr. Monro ſaw in the Engliſh army, when in Germany, that repeated ſmall doſes of this root, from four to ſix grains, occaſioned vomiting and purging; but, at the ſame time, ſuch an intolerable ſickneſs, that it was impoſſible for him to bring the ſoldiers to undergo this treatment. Dr. Francis Ruſſel found, in the year 1756, that a few grains of rhubarb mixed with each doſe of ipecacuanha, made it operate rather as a purgative; while, at the

the fame time, his patients were not fo fick with
it. Dr. Akenfide gave only one or two grains
of the root every fix hours, but took care to mix
it with a julep, compofed of mint-water and
confectio cardiaca ; and excepting breathing a
vein and giving a vomit, he feems to truft to
this medicine alone for the cure of the dyfentery.

I have myfelf, likewife, made trial of this
method of giving ipecauanha in fmall dofes, in
the epidemy of 1766, with many of my patients. I gave it to children at four times, at
each time five grains, with the fame quantity of
cremor tartari ; to adults at three times, at each
time ten grains, with half a drachm of cremor
tartari, or elfe at four times, each time ten
grains, with the fame quantity of the faid falt.
The vaft ficknefs at the ftomach, taken notice
of frequently by fo many writers, and formerly
by myfelf in two inftances, I did not now obferve ; but neither did I find, that thefe repeated
dofes procured the defired evacuations better,
than if I had given the whole at once : and many
times no vomiting at all enfued on the firft or
fecond dofe; while, at the fame time, there was
a great deal of pituitous matter in the ftomach ;
and it was only in exciting a greater evacuation

by

by ſtool, that theſe ſmall doſes ſeemed to be of any uſe.

Perhaps, however, no method is better than that, which Dr. Tiſſot has given us in the putrid fever, and which conſiſts in diſſolving a pretty good portion of emetic tartar in a great deal of water, making it palatable with ſyrups, and drinking as much of it as is ſufficient to excite repeated vomitings. That very ingenious Engliſh ſurgeon, Dr. William Ruſſel, found, during the attack of Martinico by the Engliſh troops, that the tartar emetic was the beſt and chief vomit in all caſes of the dyſentery, where there is much putrid bile in the ſtomach and bowels, as it quickly evacuates the corrupted matter, which otherwiſe did the patients in that hoſpital the greateſt harm; if it remained only for a ſhort time in the inteſtines. Dr. Pringle thinks it always ſerviceable, to mix a grain or two of emetic tartar with a ſcruple of ipecacuanha, by which means; without in the leaſt diminiſhing the peculiar virtues of the root; the medicine becomes more purgative, and more effectual in evacuating the gall. This mixture may, therefore, be made uſe of in the beginning of the dyſentery, by thoſe, who do not chuſe to order the emetic tartar alone.

The

The ſtimulus of the emetic tartar is ſo much
the more neceſſary to be added to the ipecacuanha; as this laſt has ſometimes, even in large
doſes, no effect upon ſtomachs, that are not ſufficiently irritable, or loaded with too great a quantity of mucus; while, at the ſame time, it operates very well, even in ſmall doſes, in oppoſite
circumſtances. I was called, during the epidemy at Brugg, in the year 1766, to a child
of twelve years, that had had the dyſentery
three days, with a very bitter taſte in the mouth,
a vaſt oppreſſion on the ſtomach, violent gripings, and a high fever: I ordered him for that
night, half an ounce of cremor tartari, and for
the following day, thirty grains of ipecacuanha
for a vomit; it did not excite any vomiting,
but, on the other hand, evacuated by ſtool a
great deal of an intolerably fœtid matter, with
much relief to the patient: in the evening, and
during the whole night, I gave him ſtill more of
the cremor tartari, and the next morning four
ounces of tamarinds to procure an evacuation,
which they did not immediately produce; but at
firſt occaſioned a prodigious vomiting of a very
copious, tough mucus; however, at laſt they
purged him pretty ſmartly, and the oppreſſion on
the breaſt and the pains in the belly, vaniſhed
together with the fever: the cream of tartar com-
pleted the cure. In this caſe I ought, without
doubt,

doubt, to have given the tartar emetic either alone, or with the ipecacuanha.

The repetition of the emetic is, in certain cafes, of no fmall importance. In very obfti-nate ones, Dr. Monro faw emetics, properly ad-miniftered, extremely conducive towards the cure ; and many phyficians rely on ipecacuanha alone for the cure of this diforder. In the epi-demy of 1766, I remarked, that partly on ac-count of heating remedies having been taken at the beginning, partly on account of the pre-fence of a great quantity of bilious and mucous matter, and partly by reafon of the patient's be-ing peftered with worms, which crept up out of their ftomachs, all the medicines they took were rendered ufelefs, by a continual inclination to vomit, and frequently by abfolute vomitings, in fpite of the emetics, that were given them for many days together. In thefe circumftances, I ordered very often the tincture of rhubarb made with water, which, for the moft part, ftaid on their ftomachs, and at length (though after fome time) brought the diftemper happily to an end ; but in vehement, dangerous, and urgent cafes, in which there was no inflammation, I gave a fecond vomit. A man of thirty-four years, at Brugg, was violently feized with the dyfentery : the firft day, fomebody prefcribed him

him a vomit, and in the evening some cream of
tartar. I was then called, and gave him, the
second day, tamarinds in the morning, and during the night, cream of tartar with chamomile-
tea ; the third day, manna with sal catharticum
amarum, and in the night, at proper intervals,
tamarinds ; he cast all these remedies up again,
and with them an astonishing quantity of bilious
matter; his stools were at the same time extremely frequent, but very small in quantity,
perfectly bilious, and mixed with a great deal of
blood : as well after, as before the stools, he
had continual gripings, and his fever daily increased, after having been almost imperceptible
in the beginning. Early on the fourth day I
was told, that I should not be able to give him
any more medicines, as the whole preceding
night, and that morning, he had again brought
up a vast quantity of bilious matter, and as to
the rest, found himself in the same miserable
condition as before. This induced me to try
the ipecacuanha ; my patient took it willingly,
and soon after vomited a very great quantity of
bilious and mucous matter, together with a large
worm : immediately on that, I gave a purge of
manna and sal catharticum amarum ; he did not
bring it up again, but had from it twelve large
and copious stools; the result of which was, that
his pains diminished at each stool, and, at

O   length

length entirely ceafed, without returning any more. The remainder of the hiftory of this remarkable cafe I intend to relate, when we come upon the article of the regimen of the mind.

The purge may be given a couple of hours after the firft emetic, or the morning following its exhibition ; but the ufe of purges, and the time when it is proper to repeat them, muft be regulated according to the influence they have on the pains: On this article I have again the affiftance of two of the greateft phyficians in England : his Britannic Majefty's phyfician in ordinary, Dr. Pringle ; and Dr. Monro, formerly phyfician to the Englifh army in Germany, and now phyfician to St. George's hofpital in London : they have both feen the fame things, as I have, and have both drawn from what they have feen the fame conclufions, as I. According to Dr. Pringle's opinion, whether the emetic be repeated or no, the purge muft, at all events, always be given the next day, or the day after ; and after that, as often as is confiftent with the patient's ftrength, or the obftinacy of the fever requires. However, the neceffity for continuing the purges, is rather determined by the pertinacity of the gripings and bearing down, than by the blood, that appears in the ftools ; and Dr. Pringle thinks it impof-

fible

fible to effect a cure, without copious evacuations of this kind. He admonifhes us,
therefore, not fo much to attend to the dofe,
as to the effects, which are not to be judged of
by the number, but by the largenefs of the
ftools, and the relief which the patient finds
with regard to the gripings and tenefmus after
the operation ; for the ftools are commonly
more frequent from the diforder alone, than
from the purges exhibited. In like manner, my
friend Monro found, that a great part of the
cure depended on the frequent repetition of the
gentle purges he gave in the beginning, which
are capable of evacuating the corrupted matter :
he gave purges of this kind to the foldiers of the
Englifh army in Germany every fecond, third,
or fourth day, according as the cafe required ;
for the effect of the firft purge, and the fymptoms that appeared after its exhibition, determined the number of times it ought to be repeated. Dr. Monro was aftonifhed at the little
lofs of ftrength, which his patients fuffered on
being fo frequently purged : he gave ftrong
people thefe purges, fometimes two, three, or
four days, one after another; and obferved,
that the patient, inftead of being weak upon it,
appeared ftronger, · chearfuller, and brifker,
from the relief, that enfued at the end of the operation of each purge ; as his corrupt, putrid hu-

mours,

mours, were now evacuated, that as long as
they remained in the inteftines, kept up in him a
conftant ficknefs and naufea. From thefe experiments, fo very like my own, the truth of this
great medical maxim appears, with refpect to
purges in the putrid dyfentery; that in it, no
other medicines ftrengthen the patient, than
fuch as diminifh his diforder; and that he is
very often ftrengthened moft, when he thinks
he is moft debilitated.

By the fpeedinefs of this method, in evacuating the corrupt, bilious matter, one may cut off
the diforder at the root, if there be not invincible obftacles in the way; while, on the contrary, by neglecting this method, it will, at
leaft, be very tedious. In the epidemy in 1766,
I faw many perfons cured in two or three days
by evacuations, made immediately at the firft
attack of the diforder, and fufficiently repeated;
though they had all the tokens of a real dyfentery; a violent and long-continued cold fit,
vaft ficknefs, an inclination to vomit, bitternefs
in the mouth, heat, head-ach, great pains
about the fpine of the back, gripings in the
belly, and a loofenefs, with a very fmall quantity of excrement. A woman of thirty-nine
years, at Brugg, found herfelf precifely in this
condition: I gave her, on the firft attack, in the
evening,

evening, four drachms of cremor tartari ; this
occafioned four large ftools in the night-time :
the next morning I gave her three ounces of
tamarinds ; upon this followed a great number
of large ftools, to the great eafe of the patient,
and the fever went off : I gave her, for that
night, an ounce of cremor tartari, to be mixed
with two quarts of barley-water ; and on the
third day, two ounces of manna with fix
drachms of the Sedliz purging falts, which the
fame day put an end to the diforder. I have feen
not a few examples of this kind : fometimes the
attacks were fmarter, and were neverthelefs pretty
quickly cured by the very fame method.   A
young woman of twenty-three years of age, at
Brugg, had, during the epidemy of 1766,
very violent gripings for a whole week together : and at length a confirmed and extremely
painful dyfentery, with fuch a fever as I never
yet faw at the firft attack ; a fire-red face, a
ftrong pulfe, profufe fweats, an extreme bitternefs in the mouth, and a conftant inclination to
vomit : I gave her, at five o'clock in the evening, a vomit compofed of forty grains of ipecacuanha, and twenty of cremor tartari ; and two
hours after, a drachm of the fame falt, with a
drachm of rhubarb : and it was aftonifhing,
how much bilious matter was evacuated upwards and downwards with relief to the patient.

O 3   The

The fecond day, I gave her an ounce and a half of the Sedliz purging falts early in the morning: the evacuation was copious, and the matter evacuated red and green; her pains were much relieved about noon, the fever at the fame pitch, but the heat was, however, lefs. On the third day, I gave her again an ounce and a half of Sedliz falts early in the morning; this occafioned half an hour afterwards a copious vomiting of bile, notwithftanding which, enfued a violent purging: twelve hours after that, all her pains vanifhed, and at four o'clock in the evening I did not obferve fhe had any more fever; towards night her pain came on again, I gave half an ounce of cream of tartar with a quantity of barley-water, to be drank by degrees in the night-time; this occafioned many more ftools, and her pain went off. The fourth day I found my patient in a very good condition, and prefcribed her the tincture of rhubarb; in the evening fhe had fome pain, and a bitternefs in her mouth; I ordered her fome cremor tartari for the night. The fifth day the girl found herfelf perfectly well, voided a worm, had no ftools the whole day throughout, and got quite well. But when the beft remedies, exhibited to evacuate the bilious matter in this fpecies of dyfentery, do not perform their office; the fame thing happens, as is obferved in the ufe

of

of their oppofites, namely, aftringent and conftipating medicines. In the epidemy of 1766,
at Brugg, a child of fix years, naturally bound
in his body, was feized with a dyfentery; the
bilious matter it caft up the firft and fecond
day in great quantities, convinced me, that this
cafe was of the bilious kind; the foftnefs of the
pulfe, and the pains being fo tolerable through
the whole courfe of the diforder, as never even
to make the child cry, was a proof to me of the
abfence of inflammation; the child had ftools
without number night and day, but not one
good; and the dyfenteric matter remained fo
pertinacioufly in his body, that I made ufe of
ipecacuanha, manna with a fmall quantity of
cremor tartari, tamarinds, and tincture of rhubarb in the quantity it is ufually given to children of that age, without any effect: for on the
fourth night he became perfectly light-headed,
voided a worm, and had many convulfive twitchings. I faw him early the next morning, when he
was quite out of his fenfes; his eyes were perfectly
convulfed; he toffed and tumbled about in his
bed; I felt the cold hand of death upon him, and
accordingly he died the fame day; thus giving me
a proof, that a child may very quickly die even
of the bilious dyfentery, without a violent pain
in the bowels preceding, a hard pulfe or diftended
belly, and confequently without any preced-

ing

ing inflammation, when a copious bilious matter, (that ftimulates the bowels to deadly convulfions, remains in the body. In fine, during
the above-mentioned epidemy, I faw in the moft
decifive manner, in the cafe of a reverend clergyman, how much the invincible inclination for
ftrengthening and heating remedies, and (which
is the confequence of it) an averfion for every
thing that evacuates, increafes the diforder, and
makes it more ferious and dangerous in all refpects; fo that even fo late as towards the end
of it, tumours appear on the legs, and the com-
pletion of the cure is prolonged above five
weeks; fuppofing one can force as many evacuating remedies on the patient, as is juft fufficient to fave him from impending death.

With regard to the choice, that ought to be
made of the purges fit to evacuate fpeedily the
bilious matter, Dr. Monro, Dr. Brocklefby,
and Dr. Ruffel have made various experiments,
which agree with mine. The purge which Dr.
Monro chiefly made ufe of for his firft patients,
was rhubarb; but after repeated trials he found,
in the fame manner as Dr. Brocklefby, that in
general, rhubarb during the firft period of the
difeafe, was not fo good as the fal catharticum
amarum with manna and oil, which operated
without pain or anxiety, evacuated much bet-

ter,

ter, and gave more relief, than any other purge made ufe of in the Englifh army. According to my lateft experiments, all this is perfectly true ; I gave in the epidemy of 1766, manna and Sedliz falts together in a draught, with much better fuccefs than tamarinds, but left the oil out. Dr. Monro prefcribed likewife at Bremen, the tincture of rhubarb made with water, and found it an eafy purge, but not fo well coinciding with his views in recent diforders, as the falts and manna. This is alfo perfectly juft ; but yet I faw in the fame epidemy fome cafes, in which tamarinds, manna, and falts, would not keep on the patient's ftomach ; fo that in thefe cafes, the tincture of rhubarb given in great quantities had fomething very excellent in it, as the ftomach bears it very well, as it often takes off the vomiting, and, as it at laft puts an end to the diforder, though not fo foon as the tamarinds, manna and falts ; nay it fometimes even puts the ftomach in a condition to bear the above-mentioned remedies. However, after all, thefe laft are moft fpeedy in the cure ; the cafe muft be then, that children have mach acidity in their primæ viæ, and efpecially in their ftomach, which is a great obftacle to the purging quality of the tamarinds and cremor tartari, and on that account thefe medicines are very often of no ufe at all to children. Dr.

Francis

Francis Ruffel faw at Gibraltar, in the year 1756, a rife and mortal dyfentery; after having made trial of a great number of remedies, he found that nothing gave greater relief or more forwarded the cure, than repeated dofes of the fal catharticum amarum. I have alfo made ufe of this falt with benefit.

It has neverthelefs been always thought that falts, and even acids of all kinds abraded the inteftines. It is true, that all rough and irritating remedies fhould be avoided in this diforder; but the point is to know rightly, what medicines have this effect in the bilious dyfentery, for in this very point many phyficians have been deceived. Zacutus the Portuguefe, indeed, was not afraid of arfenic in the dyfentery; but his countryman Amatus condemns even tamarinds, on account of their ftimulating acidity. Degner fays that all falts, for example, tartarus vitriolatus, arcanum duplicatum, fal polychreft, fal prunellæ, and others, are often prefcribed by phyficians, abfolutely in oppofition to all the dictates of common fenfe and medical prudence; as by their corroding properties they only exeite more violent pain and irritation in the ulcerated bowels: he did not therefore even think nitre fafe, in the height, and during the progrefs of the diforder; as it increafes the
diarrhœa.

diarrhœa. Now it appears to me, that Dr.
Degner has here drawn conclufions without fufficient foundation. For firft, it is falfe that the
inteftines are ulcerated in the dyfentery, fo often
as is in general believed; and where they fhould
chance to be inflamed, or even abfolutely ulcerated, no phyfician of any fenfe would be
venturefome enough to prefcribe a falt: in the
fecond place, it anfwers precifely the phyfician's
intention, when by means of a falt properly
chofen the flux is increafed, as long as there is any
corrupt bilious matter to evacuate. However,
Dr. Degner was not utterly ignorant of the falutary influence of acids in general; for he
gives whey great encomiums, and even com-
mends the copious ufe of lemon-juice, which he
did not find ftimulating; and is very fond of
Mofelle and Rhenifh wines, purely on account of
their acidity. Had this famous burgh-mafter
of Nimeguen, properly diftinguifhed the bilious
dyfentery from the malignant, he would not
perhaps have rejected in the bilious dyfentery,
what he undoubtedly found hurtful in the malignant.

With regard to the ufe of acids in the dyfentery, the force of truth, even in former times,
broke now and then through the clouds of prejudice. Dolæus, a writer of experience, that
according

according to the error of the age he lived in, afcribed the caufe of the dyfentery to an acid, had, however, fincerity enough to recommend ftrongly a mixture of lemon-juice and oil, and confeffes, that he had cured with this medicine above a hundred people of the dyfentery. Riverius, in all fluxes that proceeded from a putridity of the juices, ordered bifcuits to be repeatedly dipped in vinegar, then to be dried, rubbed to powder, and a foup to be made of them. Among the modern phyficians the famous La Mettrie, with others, found vinegar, lemonade, and whey, of great fervice in the common putrid dyfentery, as well as that the dread of fruit was without any grounds. Perhaps I have made more ufe of acid falts in the dyfentery, than any other phyfician; for Dr. Tiffot does not order, as it is inferted in the German tranflation of his Advice to the People by an error of the prefs, an ounce of cremor tartari, to be taken with two quarts of barley-water, but only two drachms; though at prefent he gives even an ounce at two or three times in a very little while. By this procedure I found, that the oppofition of phyficians (at leaft to acid falts) in the fpecies of dyfentery here treated of, comes entirely from prejudice.

Syden-

Sydenham has in the ſtrongeſt manner recommended by his authority, opium and its
preparations in the dyſentery; though many
objections have been made ſince thoſe times
againſt them, objections, that are certainly neither diminiſhed in number, nor otherwiſe weakened by my experiments. I will not tire my
reader with repeating them, but recommend to
him the cautionary rules, that may be drawn
from them, and which, when we make uſe of
opium and other things of that ſort, ſhould always be preſent before our eyes. Alexander
Tralles looks upon the conduct of thoſe, that
in the dyſentery, immediately pour a great
quantity of opium into the body, as raſh, and
without judgment; and Freind remarks upon
this place, that theſe remedies do indeed for a
time put a ſtop to the alvine flux, but after that
only increaſe it, and beſides this, attack the
head, and weaken the patient; upon which account Alexander is of opinion, that opium
ſhould be only uſed on the moſt extreme ne-
ceſſity in this diſorder. Soporifics and opiates
were very much ſuſpected by Degner in this diſorder; as in his opinion, it is neceſſary to be
particularly attentive to the patient during their
uſe; and as they muſt never be preſcribed without great caution, and even then not in the day-
time; that by their obſcuring and torpifying
the

the senses and vis vitæ, the malady may not lie
concealed, and thus have an opportunity to increase, and make greater devaftations. Pringle
fays, that all preparations of opium and aftringent medicines, are of fervice for a fhort time
only, and make the diforder towards the end
more dangerous than before ; it would be therefore better, that opiates were not given at all, till
the primæ viæ were cleared ; for though they
at firft give fome relief, yet by confining the
wind and the corrupted humours, they tend to
fix the caufe of the diftemper; whence it is, that
the premature ufe of opium in the dyfentery
fometimes occafions a real tympany.   This fix-
ing of the morbific caufe by means of opium,
Dr. Pringle has proved by repeated experiments ; though Sydenham does not feem to ap-
prehend much danger in it.   To be fure he did
not omit to purge his patients when the dyfentery was moft epidemic ; but at all other times
he feems to have trufted to laudanum alone.
But whatever was the nature of thefe fluxes,
Pringle was very fure that fuch as are moft in-
cident to an army, are not of fo mild a nature,
and can never be cured without evacuations; the
beft rule therefore that this excellent phyfician can
give, is to defer the adminiftration of opium, till
the patient has undergone fufficient evacuations,
and then if it be ftill neceffary, to begin with fmall
dofes.

doses. But when opium given with thefe cautions does not procure any eafe, Dr. Pringle with the greateft reafon looks on it as a token, that fome corrupt humours ftill remain in the inteftines, and that it would be of more ſervice to go on with the evacuations, than to ftop the flux.

Phyficians of equal authority are of the fame opinion with regard to opium. Eller has in like manner found, that notwithftanding the fhort relief perceived on the ufe of this medicine, the pain attacks the patient afterwards with new force; and that from its property of diminifhing the tenfion of the fibres in the inteftines, the expulfion and evacuation of the acrimony adhering to them, would be put a ftop to; and thus the difeafe, that we are ſtriving to get the better of by means of opium, is increafed. Notwithftanding this, Eller gave fome flight preparation of this remedy, but not before he had very much diminifhed and nearly put an end to the pains, and had evacuated the greateft part of the dyfenteric matter; he had however immediate recourfe again to a purge, and repeated it, as often as the tormina returned, and confequently as often as there were any figns of acrimony being collected afrefh; fhewing by this procedure, how very differently a
phyfician,

phyſician ſhould conduct himſelf in the various periods of this alarming diſorder. Dr. Young in Scotland, who has written excellently well on the virtues of opium, gave it in the dyſentery, only when the diſorder was very mild, or when its violence had already remitted by the means of evacuants and emollients. Dr. Baker in England, found opium not adviſeable in this diſorder, till the excrements were pretty nearly of their natural conſiſtence. My friend Monro found in the Engliſh army in Germany, that diaſcordium, philonium, and other remedies of this kind, put too great a ſtop to the alvine flux, occaſioned a violent cholic, and increaſed the fever; he therefore made uſe of them very ſeldom in the firſt period of the diſtemper; he gave however an opiate at night, when he had purged the patient in the day-time, and even re-peated it every night, though he had not purged him the ſame day; but found himſelf obliged to uſe a great deal of circumſpection with regard to the doſe, as long as the diſtemper remained at its height; he gave his medicines too only in a quantity ſufficient to alleviate the pain, and pro-cure ſome reſt, but never enough to ſtupify the patient, or ſtop the alvine flux, while it kept within proper limits. All theſe prudential rules coincide with thoſe, I have learned from my own experience.

Next

Next to opiates, Dr. Pringle, as well as myfelf, found nothing better for alleviating the pain, than fomentations of the abdomen and drinking chamomile-tea; the latter is alfo ufeful on account of its antifeptic quality. During the epidemy of 1766, I faw likewife at times in difficult cafes, the gripings and even the tenefmus happily quieted by drinking plentifully of almond-emulfion, and fleep promoted by the fame means. When the pains of the abdomen were too much fixed to ceafe on the application of fomentations or demulcents, Dr. Pringle applied with great fuccefs a blifter-plaifter to the part affected. Dr. Eller found a thick gruel made of barley, oatmeal, or rice mixed with a good deal of oil in a clyfter, very ferviceable in a violent bearing down of the anus; but I have likewife found remedies of this kind ufelefs, and have therefore confidered the real nature of this tenefmus, felected other medicines for it, and obtained my end in the year 1765, in the manner I have related in the fourth chapter. In the year 1766, proceeding on the fame principle in a very obftinate cafe, (in which however the patient had undergone a great many evacuations, as well at the beginning as during the courfe of his illnefs) I firft gave tamarinds for a violent tenefmus, that drove the patient to

P   the

the heighth of defpair, and afterwards the
tincture of rhubarb in great plenty, with a
good many clyfters of gum arabic, a deal of
almond-emulfion, barley-water, chamomile-tea,
and likewife a fmall quantity of opium, without that relief enfuing that might have been
expected; on the 15th and 17th day of the
diforder, I gave him a draught compofed of
manna and Sedliz falts; this procured, with
much relief to the patient, (though the ftools
were few in number) an amazingly copious
evacuation of a matter, which was at firft without fmell, and yellow, but foon after extremely
offenfive, and almoft black.   From this theory,
founded entirely on experience, I can conceive the reafon, why befides clyfters, rhubarb
is made ufe of for a tenefmus in the Eaft-Indies.

In cafe that the patient was fuddenly feized
with a terrible pain in the bowels, and a
violent tenefmus on a day, when he had
not taken phyfick,   Dr. Monro prefcribed
the purging falts with manna: but when
purges and gentle opiates were of no effect,
he ordered the whole abdomen to be covered with warm fomentations; and warm barley-water, thin rice-gruel, weak broth, or chamomile-tea to be drank plentifully by the patient; afterwards he gave emollient clyfters in
great quantities, and when they were not fufficient,

cient, a fmall quantity of the fame, with an
addition of one or two drachms of the tinctura
thebaica. For he had obferved that clyfters
of this kind, impregnated with opium,
often gave greater eafe than opium given in
any other manner. When the bearing down
in particular was very troublefome, a clyfter
compofed of ten ounces of water, one ounce
of the mucilage of gum arabic, two ounces of
oil of olives, with a proper quantity of diafcordium and tinctura thebaica, or elfe a ftarch
clyfter, gaye more relief than any thing elfe.
In fome cafes, in which the pain was quite intolerable and accompanied with fever, the
Doctor found himfelf obliged to breathe a vein,
and fometimes to lay a blifter on that part
of the abdomen, where the patient felt moft
pain.

Blifters not only act as palliatives in the dyfentery, but likewife contribute towards the
cure; being of the utmoft ufe in extraordinary
cafes of this diforder, as well as in the immode-
rate diarrhœas attendant on putrid fevers, and
indeed in general in all obftinate alvine fluxes.
During the epidemy of 1766, I faw fome flight
dyfenteries in children of divers ages, from a
year and a year and a half, to feven or eleven
years old, extraordinary obftinate, and fome-

times

times very tedious. My friend Tiffot faw fomething like it about the fame time, and complained likewife of the fame obftinacy and tediouſneſs of the diforder, but called it only a loofeneſs; I on the contrary, by what I could fee of it, judged it to be in reality nothing elfe than a flight dyfentery, as it was accompanied with a conftant and fometimes very violent fever, which was full as pertinacious as the flux: however, our care was not about the name, but about the beft method of cure; and this was found out by Dr. Tiffot. The children under my care had moft of the fymptoms attending the putrid dyfentery, though they did not come on them on a fudden: I fometimes faw an aftonifhing quantity of tough, thick mucus, vomited up at repeated times; and yet they had from forty to fixty ftools in twenty-four hours, which were frequently very bloody, and of all forts of colours, and always very fmall in quantity; they had however much lefs pain than is commonly obferved in the dyfentery, and for the moft part none at all: one child only had a prolapfus ani. With fome I made ufe of an emetic in the beginning; with others, of tamarinds, and with all, of the tincture of rhubarb and chamomile-tea: in this manner I cured in twelve days time, a boy of nine years of age, that for feveral years paft

had

had been in a perfect confumption, and afflicted
with various ulcers; he had indeed been fomething better for a few months before, but was
ftill plagued with a confiderable ulcer about
the fpina dorfi, and a conftant hectic fever. I
had not the like fuccefs with all the children
under my care: fome did not get well in lefs
than three weeks; and one, notwithftanding all
the pains I took, and the application of three
blifters, not in lefs than a month; though to
be fure the blifters were applied too late, and at
a time when the fever was at a very high pitch,
and the abdomen ftretched as tight as a drum.
Notwithftanding thefe misfortunes, veficatories
merit the preference above all the remedies
hitherto mentioned in thefe obftinate cafes. Dr.
Tiffot ordered them to be laid on eleven children: on one they had no effect; with another
they had a vifible, but tranfitory fuccefs; with
all the reft they did more towards the cure,
than all the other remedies employed for that
purpofe, and they cured children that could
not otherwife be brought to take any medicine
at all. The Doctor commonly ordered them
to be applied to the calves of the legs, and,
when the belly was diftended, to the nape of
the neck; but I, for my own part, in this
cafe, laid them upon all three places at
once.

P 3   With

With refpect to the diet in this fpecies of
the dyfentery, I have already treated of the
moft neceffary articles in the third chapter;
however, there remain a couple of remarks
and obfervations, which perhaps are not abfolutely fuperfluous. All hard and indigeftible
food occafions in this diforder pernicious ob-
ftructions, as the bowels which are extremely
weakened, and indeed almoft difabled, are not
capable of propelling fuch a globular mafs as
muft here be neceffarily formed; I cannot
therefore comprehend the reafon, why Degner,
during the whole courfe of the dyfentery at
Nimeguen, allowed his patients to eat pota-
toes. On the other hand I now perceive, why
fome phyficians in the foregoing century,
fhewed fuch an averfion to drink; as I have
fince that feen cafes, in which every fpoonful
the patients eat, 'if they drank after it, gave
them a ftool on the fpot; but this very thing
ought to have been an indication to them,
to allow their patients to drink the more,
inftead of forbidding them to drink at all.
Monro prefcribed for the foldiers under his care,
barley-water and thin rice-gruel in great plenty; and nothing, according to this great phyfician's obfervations, conduced fo much to the
cure of the dyfentery, as the copious ufe of
fuch drinks, as obtund and involve the acrid
humours,

humours. In the epidemy of 1766, .I heard much faid in praife of milk, efpecially in various parts of the canton of Zurich; but I did not hear of any genuine and good obfervations, as nothing but the common talk came to my ears. Dr. Pringle never allowed his patients milk, even when they were getting well, except it was diluted with lime-water; as he obferved, that milk by itfelf had a great tendency to increafe the gripings. I allowed many of my patients grapes in the above-mentioned epidemy, without having ever obferved any thing in them, otherwife than what was perfectly innocent; but, on the other hand, found in fome obftinate cafes, when the patient was recovering, though very flowly, that this fruit being permitted him without laying afide his other medicines, at firft purged him very well, afterwards diminifhed the ftools by degrees, and finally reftored him to perfect health.

The mind, in the dyfentery, as in all other diforders, has likewife need of a regimen. I fhall in this refpect only cite a couple of obfervations, to fhew how the mind may do harm to the body in the dyfentery : for more refined and metaphyfical confiderations on the regimen for the mind, a fubject, I intend to treat of in

P 4   another

another.work, would not fuit with the fimplicity I have adopted in thefe pages.

The firft obfervation is on the pernicious effects of impatience.   I have already faid in my
work On Experience in Medicine, that fuch
perfons as will not bear with any thing, are not
only fooner ill than others, but are alfo much
longer and more violently fo ; becaufe the continual alarum of their paffions, their torment-
ing fears at every increafe of their indifpofition,
their reftlefs and unremitting anxiety, that is
not to be foothed by the tendereft endeavours
of their neareft and beft friends, always recalls
to them the fenfation of their malady; and likewife becaufe by the perpetual diforder of
their fenfitive faculty, the regular and ufual
courfe of the diftemper is changed and dif-
turbed.

A man of forty-five years, at Brugg, that
had been more ufed to indulge himfelf in the
pleafures of life, than bear with its calamities, who
for that reafon had already been almoft driven
to defpair with a mere head-ach, and in many
confiderable illneffes had been a melancholy
inftance of the greateft want of fubmiffion to
his fate, that is poffible to be fhewn by man,
was attacked in the epidemy of 1766, with a
violent

violent dyfentery. He had for a long time the advice of our famous and admired phyfician Dr. Fuchflin, and at laft, without the leaft neceffity for it, called me in to confult with him. I fearched into the circumftances of the patient's complaint, with the greateft attention, and found that Dr. Fuchflin had purfued the beft method poffible: all the medicines he prefcribed, had been effectual; the fever was cured, his gripes were gone, and the colour of his excrements were natural, only his ftools and tenefmus were ftill confiderable; in fhort, I found the fick man happily delivered from the danger of death by his firft phyfician, the illnefs quite on the decreafe, and nothing now feemed more to be done than to finifh the cure. This province was given over to me. I laid it down as an indication, to remove the tenefmus gradually, by attemperating and expelling the peccant matter ftill remaining in the cavities of the inteftina craffa, and fo by this means put an end to the diforder. The patient had much eafe; often during a whole day or night very little bearing down, and a good fleep of many hours duration: but yet the tenefmus returned from time to time, and each time in confequence a ftate of defpair, which it is impoffible to defcribe or give a name to; the patient's mind was obfcured with a dark

cloud,

cloud, which vanifhed as foon, as he had company. But when his friends took their leave, all comfort, hope, or views of future joy went with them; his ufual anxiety then returned, with its long, black, and horrible train of melancholy ideas, even when he had no pain, even when he had juft awaken out of his fleep and found himfelf in bed. The effects of this terrible condition of his mind, were now very bad, and therefore deferve to be defcribed in a work, that Is intended for the benefit of mankind. This truly to be pitied man, had been as well purged at repeated times of all acrid and corrupt bilious matter, as he poffibly could be; his excrements very often did not fhew the leaft figns of corruption, but in confequence of his miferable lamentations, whining, and mortal anxiety at each twitching about his anus, his bile overflowed each time, and immediately on that his excrements were green. I was thus driven backwards and forwards, without coming nearer to the point for many days, till at length, by the ufe of fome medicines opportunely exhibited, thefe twitchings about the anus went off: but it was five weeks, reckoning from the commencement of his diforder, ere he grew perfectly well.

The

The fecond obfervation is on the violent effects of wrath. Not the vulgar, but people of penetration, that is, heads that are capable of comprehending things in a philofophical manner, will eafily fee from this account, that a violent and often repeated provocation to anger, in a dyfentery otherwife attended with a putrid fever, had converted this putrid fever to a malignant one, and that a fever of this fort became mortal, from a difpofition of the bile quite different from that in the preceding cafe; though there was no inflammation in the bowels, and therefore nothing that could occafion a gangrene in thofe parts.

A man at Brugg, under the middle age, who by nature was inclined to anger, and was in circumftances of being often provoked to effufions of the gall, and had befides for many years before been very often feized with a fudden ficknefs at the ftomach, had during the epidemy in 1766, the dyfentery for four days, in the fhocking manner defcribed above under the article of Emetics. On the fifth day in the morning, he brought up fix large round worms, he was as to the reft, ftill free from his pain, but not from his fever; the very fame evening fix more round worms came again out of his mouth; in the night he was obliged to go go very often to ftool, his excrements were now white, with but little blood in them, and the patient himfelf free from pain. On the fixth day he had a good many ftools, but was ftill without pain; on the feventh day, his ftools were as few again as on the foregoing, his fever was very inconfiderable, and he without pain; he was juft in the fame ftate during the whole night.

On the eighth day at five o'clock in the evening. I perceived on coming into the fick man's room, a horrid change. His face was as pale as death, his lips white, his eyes fixed and yellow, his looks wild, his whole body in anxious motion, and all his difcourfe was nothing elfe than a continual very boifterous clamour for cold water. Aftonifhed in the higheft degree at the horrid afpect of my patient and his diforder, I however afked the poor man with the greateft feeming indifference, if he had not had any great pain in the bowels, which had gone off on a fudden? he anfwered me, No: Whether fince the fourth day of his illnefs, notwithftanding the frequency of his ftools, he had not felt any more pain in the leaft in his bowels? he anfwered me ftill in the negative: I then afked him whether he had not found a difficulty in fwallowing fince the morning? he faid he had: Whether he did not perceive a vaft bitternefs in his mouth? Yes, he did: If
he

he had not anxiety at the breaft? he had:
Whether he was obliged often to go to ftool?
Yes: Whether his excrements were black?
No, they were not: If they ftank very much?
No: Whether his urine did not fcald him? it
did. The by-ftanders told me withal, that
the fick man dozed for a couple of minutes,
at which time his eyes were convulfed, and
that at times he was perfectly light-headed;
his voice was alfo utterly changed, his pulfe
feverifh, but weak; and in fhort, he was not
to be known again. I gave him quickly fome
few tender admonitions of no fignification, then
flipped away out of his room, and told the family, that I would lofe my head, if there was
not a quite peculiar external cafe, which they
would not tell me of, that had fubverted the
courfe of my patient's diforder in fo extraordinary a manner. After a long fcrutiny into
the affair, I now firft found out, that he had
been often vexed in the courfe of his diftemper, but had that day in particular had a violent
fit of anger, which had immediately brought
about the change; I now comprehended every
thing. Agreeable to the ufual effects of wrath,
the patient was farther attacked in the night-time
with a very bad pleurify, a confiderable cough,
and a violent head-ach; the anxiety too at his
breaft

breaſt remained, together with a vaſt bitterneſs in his mouth, and he had every hour three
ſtools, which were ſtill partly bloody. I thus
ſaw a great effuſion of the bile, and real ſymptoms of malignity, joined with a very violent
dyſentery.

On the ninth day early, I found the patient's countenance as pale as before, and the
whites of his eyes quite yellow ; but his looks
were not quite ſo wild, and his lips a little red-
der. I had it now in my power, with ſome
rays of hope, to bring his extremely dejected
mind into a perfectly quiet and gentle ſtate ; I
therefore, notwithſtanding the vaſt danger he
was in, ſtill gave him hopes of life, and every
time I came into his chamber, put on an air
of chearfulneſs. After I had now begun to
treat the diſorder as a malignant one, my patient vomited with great relief. He had no
more ſickneſs at ſtomach, his pleuriſy vaniſhed, the bitterneſs in his mouth went off, his
head-ach was very tolerable, he had a better
complexion, and all this happened in ſo ſhort
a ſpace of time, as from the morning till the
noon. At five in the evening his complexion
was again very good, his eyes were no more
yellow, and his head-ach had ceaſed ; but all
day long he had five or ſix ſtools every hour.
I looked

I looked at them, and found them of a citron-yellow colour, very fpumous, fomewhat mixed with blood, and yet not offenfive to the fmell. He ftill complained of a very violent heat in his urine, and even of an excoriation of the extremity of the urethra, alfo of an oppreffion and ftrangulation about the region of the ftomach, and of an inclination to vomit; all the night long he had fix ftools every hour, very fmall in quantity, red, yellow, and green. As to pain, he had none in the leaft, but was very weak.

Early on the 10th day, I found my patient to all appearance quite without fever, but ftill plagued with the oppreffion and ftrangulation at his ftomach, and to be fure infinitely weak. Soon afterwards he vomited a copious thin matter as green as grafs, with three large and live round worms; in a moment the oppreffion and ftrangulation in his ftomach vanifhed, and he again became fomewhat chearful. All that day he had fix or feven ftools in an hour, which were yellow, green, and white. At feven in the evening, I found him without that anxiety, indeed, about the pit of the ftomach, but infinitely weak and dejected in body and mind; but on having taken a proper cordial, he faid he found himfelf heartily well, flept

likewife

4

likewife from time to time, and had, inftead
of feven, but two ftools in an hour, which
were in other refpects like the former. On the
eleventh day, I did not fee my patient till
three in the afternoon, and immediately found
in his countenance fuch a chearfulnefs, as I had
not before remarked in him, and much more
ftrength in his voice; he had not more than
two ftools in an hour, which were lefs bloody
than the former ; his fever appeared to be very
moderate. All night long till morning, my
patient was aftonifhingly chearful, brifk, and
free from all pain.

On the twelfth day my patient was again
extremely vexed, which coft him his life. His
eyes and face were of a vaftly deep yellow colour, his looks were wild, and his mind was
quite overwhelmed with melancholy; he was
obliged to go to ftool two or three times in an
hour, he had fome fever, and great heat in his
urine, but otherwife not the leaft pain in the
abdomen, nor bearing down; in the night,
indeed, he had only two ftools every hour, but
no fleep; a confiderable degree of terror and
anxiety of mind, yet the heat of his urine
went off. The thirteenth day he had two ftools
an hour, not the leaft pain in the abdomen,
a yellow complexion, a violent cough, a confiderable

fiderable hoarfenefs, great difficulty in fwallowing, a pulfe flower than natural, and very dejected fpirits; in the night-time two ftools an
hour, (with one of which he had voided a large
round worm, the feventeenth that had come
away from him in this ficknefs) no pains in
his belly, but a conftant cough. On the fourteenth day in the morning, I found this cough
fo violent, that he could fcarcely fpeak; he was
very hoarfe, his eyes very yellow, he had to be
fure fome fpirits, but thofe very low; no pain
in the belly, no bearing down, but a vaft irritation and oppreffion at his breaft. From the
morning till noon he voided nothing but a yellow liquor without blood; from that time till
four in the evening he had very few ftools,
more oppreffion on the breaft than before, a
continual violent cough, a flow and weak pulfe,
his eyes fixed, and a very hoarfe voice. From
four till feven o'clock he had two ftools, confifting of a yellow watery matter. At feven
o'clock his voice was almoft gone; he had an
inclination to dofe, was very little fenfible,
fometimes an anfwer could be got from him,
but with inconceivable trouble; he fetched
his breath with great difficulty, his pulfe was
very weak, and fcarcely quicker than in health;
he rattled gently in the throat, his tongue was

Q   of

of a dark-brown colour, and at ten o'clock at
night he died.

Thus, by the force of the moſt powerful of
all paſſions, the putrid ſpecies of the dyſentery
turned, in this poor man's caſe, to the malignant ; and juſt as theſe ſymptoms of malignity
began to ceaſe, a new fit of paſſion brought on
once more an effuſion of the bile, and a fatal
depoſition of that humour on the breaſt. In-
ſtances of this kind are by no means uncommon.

I hope I may be permitted to farther give a few
hints on the manner, in which we muſt conduct
ourſelves ; when the cure is imperfect, when a
relapſe is apprehended, or when the patient has
actually ſuffered one. Dr. Pringle preſcribes
in the firſt caſe the ſame diet, as during the
firſt attack, and ſome gentle aſtringents; this
laſt intention he fulfilled with lime-water, of
which he gave a pint every day, with half that
quantity of boiled milk ; he ſometimes found
ſmall doſes of the bark not leſs effectual, added
to the extract of Campeachy-wood, or the tinctura Japonica : though it appears to me, that
tincture of rhubarb alone made with water
may ſerve inſtead of them all ; and Dr. Monro
has found, juſt as I did, that rhubarb towards
the end of the diſorder is of great ſervice,

though

though at the beginning it did not anfwer his
expectations. Dr. Eller advifes gentle aftringents and corroborants joined with flight opiates, towards the end of the diforder, when a
great diminution or perfect ceffation of the tormina fhew, that all the acrimony is evacuated.
Thefe remedies are cafcarilla in powder, or the
extract of it given with fimple cinnamon-water,
with an addition of the extract of orange peels,
and a fmall quantity of the pillulæ ex cynogloffo.
But on the leaft griping that his patient complained of, he had immediate recourfe with the
greateft reafon to rhubarb and manna; and laid
it down as a maxim, that evacuants of this
fort muft be repeated as often as the pain returns; that the acrimony may not be collected
again by degrees, and produce the diforder
afrefh.

In relapfes, what was done in the original
diftemper, fhould be likewife done now, in proportion to the patient's ftrength. But it muft
be remembered withal, that it is full as dangerous to fuppofe too little ftrength in the patient, as too much. In the epidemy of 1766,
I faw relapfes happen to fome children from
laying their medicines afide too foon, and likewife to fome adults, from their having expofed
themfelves too foon to a moift air, or ventured

too

too foon on indigeftible food, or from their having been put into a paffion. I cured the children with manna, tincture of rhubarb and almond-emulfion; and adults with rhubarb, or
cream of tartar, or elfe with this falt alone;
though fometimes I returned again to the ipecacuanha. A young woman at Brugg, about
thirty years old, having been expofed to
the rain during the vintage the whole day, was
wet through to the fkin, and immediately upon
that was feized with the dyfentery in an extraordinary violent manner; I gave her forty grains
of ipecacuanha, and the fame quantity of cream
of tartar, to be taken all at once, and two
hours after that, a drachm of powder of rhubarb, with the fame quantity of the above-mentioned falt for one dofe; with all thefe medicines, fhe voided both upwards and downwards
an amazing quantity of a bilious matter,
with much pain. The fecond day I gave an
ounce and a half of Sedliz falts; this likewife
brought away an aftonifhing quantity of bilious
matter, with the greateft relief to the patient;
I gave her then for that night, half an ounce
of cream of tartar to mix with a quart of
barley-water; this fhe drank all out, and her
pains went off entirely. The third day fhe
thought herfelf perfectly well, and in the morning before break of day went to work in a very

damp

damp place; but not being able to stay there
above an hour, went away, and was seized with
a violent continued cold fit of a fever, and with
such a pain in the abdomen, that she did nothing but turn and wind in her bed, with the
most horrid cries; I gave her forty grains of
ipecacuanha, with the same quantity of cream
of tartar, divided the whole into four doses, and
ordered her to take one of them every hour, with
a good deal of chamomile-tea. This occasioned
her, without any extraordinary sickness at stomach, to vomit after each dose, only once indeed, but that gave her great relief and a good
many stools; in the night-time I gave her half
an ounce of cream of tartar, to be taken in barley-water, with which my patient, after a good
many copious stools, found herself better and
better by degrees. On the fourth day in the
morning, she wanted to slip from me again, but
I forced her to stay at home and follow my orders: I let her have nothing else all day long
than almond-emulsion, her stools were no longer
copious, and her pain very inconsiderable; but
towards evening she was put into a passion, and
directly upon that her pain was three times as
violent as before. The fifth day I gave her
eighty grains of powder of rhubarb, with the
same quantity of cream of tartar, and divided

Q 3   it

it into two doses, which purged her very smartly and put an end to her complaint.

To this section on the cure of the bilious dysentery, I will yet add a sketch of an universal method of curing it, which deserves notice, and which I recommend to be made trial of, though it should not square throughout with my own opinion; for I am not at all solicitous about that, but only about the truth, which ought to be above any other consideration. Dr. Duncan, one of the physicians in ordinary to his present Britannic majesty, followed this method in the dysentery of 1762.

After having taken away a greater or less quantity of blood, according to circumstances, from such of his patients as were plethoric or feverish, he gave every half hour four ounces of the following julep, till it occasioned a vomiting and purging; namely, three grains of tartar emetic, and two ounces of manna dissolved in a pint of barley-water. The next day, and for five or six days following, he gave his patient as much as was necessary to purge him well, of a drink composed of manna, tamarinds, and tartarus solubilis. When the irritation and pain were very great, he found manna dissolved in the almond-emulsion, suffi-

cient

cient for his purpofe: if the pain and tenefmus were too violent to be borne, he found a clyfter of chicken-broth, or an infufion of linfeed in warm water, with an ounce or two of oil of fweet-almonds diffolved in the yolk of an egg, of great ufe, given once or twice a day. He was in general very well pleafed, when the firft evacuation by ftool was very large ; and he was ftill better pleafed, when he was able to procure this by gentle means. In this manner he often cured the dyfentery in a few days, without giving any more medicine. But when the diforder lafted above fix or feven days, he then added thirty or forty drops of the tinctura thebaica to the clyfters, and ordered a fcruple of extract of Campeachy-wood to be taken three times a day, in a proper kind of vehicle. He allowed his patients nothing to eat but boiled rice, fago, panada, and the like, but no meat, and not even chicken-broth in the beginning of the diforder, no more than oil, butter, or any other kind of fat. For their drink in common, he gave them almond-emulfion, thin rice-gruel, or barley-water with gum arabic. Out of eighty fick of the dyfentery, Dr. Duncan loft but one, who was dying when he fent for him ; and they were all treated in this method.

I con-

I conclude in fine, with a word or two on some remedies and methods, which are erroneous in dyfenteries of the bilious kind. All very violent emetics and purges ought to be rejected ; as by the former the body is attacked too roughly, and by the latter all its liquid parts are brought into the inteſtines, digeſtion is ſpoiled, the inteſtines weakened, and sometimes covered with ſlight exulcerations, that terminate in an incurable diarrhœa. Scammony, aloes, and all reſinous purges are improper, and increaſe the pain. Many phyſicians of great conſideration in Switzerland make uſe of nitre; as this diſorder is undeniably accompanied with a fever, and, as they imagine, every fever requires nitre. But Dr. Tiſſot has ſhewn, that nitre is of more hurt than ſervice in the putrid fever, and that it rather promotes than leſſens putrefaction ; as it only diſſolves the putrid matter more than it was before, and makes it fitter to paſs into the blood, inſtead of properly evacuating it. I therefore look upon ſalt-petre, at leaſt in the bilious dyſentery, as entirely uſeleſs ; ſince in the opinion of that great enemy to all empirics, Dr. Hirzel, (whoſe penetration and caution in examining into the cauſes and effects of every thing, that happens to his patients in the courſe of his practice,

are

are not to be exceeded) it is of no real ufe, with refpect to the diforder itfelf.

Of all the fpecies of dyfentery, the malignant deferves the greateft attention, as well of itfelf, as more efpecially on account òf the method of cure appropriated to it, fince this is alfo very con-trary to all other methods; and as Mr. Rahn's famous work on the dyfentery is ftill overcaft with fome clouds, with refpect to the diftin-guifhing character of this particular fpecies. The precife determination of this character lays a much better foundation, (at leaft in my opi-nion) towards an exact knowledge and judicious treatment of this peculiar fpecies, than mere re-ceipts; efpecially as according to Gruber's ac-count, in the epidemy of 1746, in the middle of the city of Zurich, a world of people died for no other reafon, than that many a good ho-neft practitioner with his receipts in his hand, was not acquainted with this truth, which indeed is not to be found in the code of empiricifm.

In the cure of the malignant dyfentery, a pure air is above all things requifite. It has been found in the army, that the more the fick are diftant from one another, and the better the hofpitals are provided with purer air than ufual,

the

the lefs number of people there dies of this
otherwife extremely dangerous fpecies ; fo that
thofe quarters were always the wholefomeft for
the foldiers, in which, on account of broken
windows and other want of repairs, the frefh
air could not be kept out.   In general the
greateft danger proceeds from an impure air,
which can never be made amends for either by
diet or medicine.   But cleanlinefs likewife in all
refpects is here of extreme importance ; for we
find in the military hofpitals, that not only the
moft airy and fpacious quarters muft be pitched
upon, and the number of fick in them be leffened as much as poffible, but that in general
the hofpital and patients muft be kept extraordinarily clean.   If all this be not done, the malignity fpreads to the other patients, a great
number of them die, and even the moft powerful remedies are without effect.   Nay, when the
infection once becomes confiderable, the greateft caution, and fome fpace of time is requifite,
before the hofpital is entirely quit of it.

It muft be owned that thefe obfervations have
been made in the camp, and may therefore be
thought of no ufe to my peaceful country ; but I
have already faid in my work on Experience in
Phyfic, that a worthy minifter told me, he often
went into houfes during the dreadful dyfentery of
1756,

1756, in a village fomething leis than an hour's diftance from Bern; where in one very low, fmall, and clofe chamber, a couple of dead bo-dies lay upon the table, and four or five per-fons fick of the dyfentery, men, women, and children, in their bed, who had a veffel ftanding open by them, in which they did their needs. We fee then plainly, that we have alfo cafes at home, in which the dyfentery, through the accidental corruption of the air, muft not only be contagious, but even extremely malignant; as without this additional circumftance it has influence enough to breed the malignant fever. It is therefore alfo not to be doubted, that a fever of this kind not only infinuated itfelf in the year 1750, in more than one place apper-taining to the canton of Bern, but alfo in 1749 and 1751, in which fame years the people of the canton of Bern in like manner died of the dyfentery in great numbers. Now, even in thofe epidemies of the dyfentery, which are indifputably benignant, as well as in epidemies of the malignant fever, there are always here and there malignant dyfenteries, in which the obfervation of thefe rules is of the greateft im-portance. And as we fee plainly, that malig-nant diforders invade Switzerland daily more and more, perhaps the dreadful time may come

but

but too foon, in which fuch rules will be looked upon with lefs contempt.

Evacuations muft be fometimes entirely omitted in the malignant dyfentery; fometimes emetics are noxious at the beginning, and on the contrary purgatives are ferviceable. It is very often neceffary to give a vomit at firft, and afterwards purges.

Now and then, and efpeeially in cafe the diforder be not as yet well known, a vein is opened directly at the beginning of malignant fevers; when the patient complains of a violent head-ach, and his pulfe is quick and full. Venæfection is even fometimes repeated; if a pain in the fide, or a violent one in the bowels difcovers itfelf; if the patient be otherwife of a ftrong conftitution, and if a plethora ftill be obferved at the very time; that one thinks of giving the bark; but in other circumftances it has always been feen to do harm, and weaken the patient too much. Venæfection has been advifed and tried in Switzerland without due confideration, even in the malignant dyfentery; but in other places the fick have been feen to die upon the ufe of this remedy in the moft deplorable manner; and I read in Dr. Baldinger's important treatife on the Difeafes of the Army, which,

which, to my detriment, was not known to me till juft before this fheet was printed off, that it was at leaft not ferviceable in the Pruffian army during the laft campaign. I, for my part, entirely rejeĉt venæfeĉtion in the malignant dyfentery on this account, efpecially as I have already found it unneceffary in the bilious dyfentery.

Emetics too and purges muft be omitted, when the excrements are quite watery, and fo inexpreffibly copious, that the patients within the fpace of twelve hours feem as if they were dying, and fometimes aĉtually die ; in which cafe all evacuating medicines fhould be avoided, and recourfe be immediately had to ftrengthening and conftipating medicines.

Emetics are omitted occafionally, when in particular circumftances or times experience fhews that they are noxious : and fometimes they do not make the patient vomit in the leaft. In that extremely malignant dyfentery, that happened in Saxony in the year 1746, and which is fo excellently defcribed by Dr. Vater, ipecacuanha at the beginning of the diforder was manifeftly hurtful; though on the other hand, during the progrefs of the fame, it did very good fervice. In the malignant dyfentery,

tery, that appeared the fame year, though in a
much lefs degree, at Zurich, emetics were found
juft as noxious; for as to their effects in benignant cafes, it is not here the queftion.
Siegefbeck has defcribed in the Breflaw Journals,
a dyfentery of the malignant kind, obferved by
him in the year 1717, in which the choiceft
ipecacuanha, which had excited vomiting in all
other cafes, did not provoke it in the leaft at
the beginning of the dyfentery, even given at the
time when the patient abfolutely retched to vomit. In like manner I faw in the epidemy of
1766 at Brugg, an extremely malignant cafe in
a boy of feven years, whom, on the firft day of
his diforder, I could not get to vomit with
twenty grains of the beft ipecacuanha, neither
had he any ftool from a purge, that I gave him
directly after.   On the fecond day he was obliged to go to ftool often, but, except a large
worm, he hardly voided any thing at all; he
looked very oddly, his eyes were as fixed
as if they were made of glafs, his head vaftly
heavy and ftupified, and the lad was indifferent
to every thing, even to the pain in his belly,
which he only fpoke of now and then when I
afked him about it; I could not find any pulfe
he had, though I felt for it all over his body.
About eleven o'clock at night he grew quite cold,
had often convulfive motions in his eyes, was

forced

forced to go to ſtool four or five times an hour, his excrements were black, and each ſtool did not amount to the quantity of half a tea-ſpoon-ful. The third day in the morning, I found the child in the ſame ſtupified ſtate as before, his face and lips quite pale, his eyes fixed, his tongue brown, no pulſe in any part of his body, though none of his limbs felt in the leaſt cold; he often fetched a deep ſigh, and told me with the moſt extraordinary indifference, that he had a pain in his belly. On his hands, arms, back, neck, and breaſt, I found many thouſand very ſmall brown and bluiſh ſpots, the very worſt ſort of petechiæ. I ſaw him again at two in the afternoon; he went to ſtool very often, but each ſtool was not half a tea-ſpoonful; about four o'clock he was at times cold, and the ſpots then appeared paler, he went ſeldomer to ſtool, and did not void any thing; at ſix in the even-ing I found him in the ſame condition, with the ſame indifference about his pain, which ſtill re-mained; he had a frequent and very violent bearing down, and did not bring away the leaſt excrement. From nine till eleven o'clock at night, he ſtill ſpoke at times, when he was ſpoken to, but always with the greateſt indif-ference, and without any other ſymptom inter-vening, he died at half an hour paſt one the next morning.

Where

Where then emetics do not succeed, we keep
to purges alone; and when these too, as in the
case just related, have no effect in the least, we
endeavour to promote sweat, if nature appears
to incline to this evacuation. Professor Vater
gave with no small relief to his patients, in the
dreadful epidemy in Saxony, directly at the beginning of the disorder, gentle and sometimes
likewise smart purges; the first to such as had
very copious stools, the latter to such as with a
considerable bearing down, had no stools at all:
however, he kept to gentle purges even in the
last case, when the tenesmus was quite intolerable: in all cases he gave three or four hours
afterward something strengthening, and repeated
this procedure every two days, with good success.   In the epidemy at Zurich, the judicious
Dr. Landolt, at that time chief physician to the
city, gave sudorifics immediately at the first
attack of the disorder ; when the patient had as
yet no pain nor alvine evacuation, by which
means he in effect forced out the exanthemata
with good success; but when the sick did not
ask his assistance before the fourth day; when
their pain was great, and their stools copious ;
the Doctor prescribed a dose of rhubarb, and
did not till after that provoke the sweats.   Perhaps in this last method of proceeding, the eruptions did not follow; though Dr. Gruber does

not

not fay any thing about it: but I know too
well, both from my own experience and that of
others, and will prove it in its proper place by
appofite obfervations, that in general a miliary
eruption, and even the petechiæ, may be very
often prevented, when proper evacuations are
made directly at the commencement of the
diftemper.

On the other hand, when the ftomach and
bowels are loaded with great quantities of corrupt
matter (which is often, though not always the
the cafe) the evacuation of the former is of vaft
importance, as well as of the latter; efpecially, if
the caufe of the fudden and total dejection of the
patient's ftrength be fituated in the ftomach. In
general ipecacuanha is ufed in malignant fevers,
when the patient complains of much ficknefs at
ftomach; and he commonly finds himfelf better a
few hours after having taken it. This emetic
fhould by rights be given at firft; but (when it
has happened to be neglected fo long) it has
been often found very ferviceable, when given at
the eighth, ninth, nay twentieth day in malignant fevers; and it may be always given after
the commencement of the malady, when no inflammation is obferved in any part, and the patient has ftill fome force left. It is alfo repeated
in the courfe of the diforder with good effect;

R   if

if the naufea and ficknefs at ftomach return, ᵣor
in cafe that the ftools fmell uncommonly offenfive.
Huxham faw very often in a malignant fever
an aftonifhing change for the better enfue on a
vomiting with ftool fupervening on the eighth or
ninth day. Dr. Brocklefby is convinced from
repeated trials, that gentle vomits prefcribed
even a fecond time on the feventh and eighth
day in thefe fevers are of fervice, and even after
that time, if ordered with prudence. There is
much alfo to be learned with refpect to the evacuation of the putrid humours, from what Dr. Baldinger fays on the method of cure in the fever,
with which the Pruffian foldiers were afflicted,
by the German phyficians ufually called the
malignant catarrh fever, and by me fimply the
malignant fever.

- Thefe obfervations are however not to be
abfolutely relied on without fome exception in
the malignant dyfentery. Ipecacuanha is without doubt the principal remedy in this fpecies of
dyfentery; but it is of the greateft importance
to give it by way of emetic directly at the beginning, before all the humours of the inteftines
are infected : its operation is promoted by chamomile-tea, which perhaps is much more requifite in this diforder than in any other; as.
chamomile flowers are extremely antifeptic.

Seven

Seven or eight hours after this firft evacuation by vomit, rhubarb fhould be given in order to procure an evacuation alfo by ftool.

Some great phyficians do not ftick at making ufe of manna, fal catharticum amarum, and oil, or any flight purge of that fort ; but prefcribe after it a gentle opiate at night, to eafe the patient's pain, and give him fome reft : they likewife repeat this purge three or four days after, that the putrid excrements may not be collected in the bowels in too great quantities. Dr. Monro does not hefitate to give gentle purges from time to time throughout the whole courfe of the malignant dyfentery, when he has done every thing elfe in the mean time requifite in a dyfen- tery of that kind. A perfon of equal reputation, Dr. Baldinger, late phyfician to the army of the king of Pruffia, gave purges as long as any pain was perceived ; but he alfo faw with great penetration into the true ufe of ipecacuanha, which he mixed with equal parts of rhubarb, and gave the firft day to the quantity of twenty grains, and afterwards five grains three times a day.

However it is beft perhaps, after having purged the patient with rhubarb, to have re- courfe again to the ipecacuanha alone, and that on

account

account of the antifeptic powers quite peculiar to it, and of the great ufe, it is found by experience to be of in malignant diforders in general. This however muft be given only in very fmall dofes to the quantity of two, three, or four grains at moft, every two hours, with a tea·cup full of chicken-broth, or of veal-broth, to which is added a fmall quantity of chicken-broth, with fcorzonera roots, carrots, or celery.

Thefe broths (as much as I have otherwife fpoken againft them in the bilious dyfentery) fhould be the patient's only food, and that with a view to keep up his ftrength: for the degeneracy of the humours in malignant fevers feems to differ from the degeneracy of the fame in the bilious fpecies not only in degree, but even in charac-teriftic fymptoms: the confiderable difference fubfifting between the medicines employed in bilious and malignant dyfenteries, fhews the ne-ceffity for making this difference in diet, efpe-cially as it is of great importance to keep up the patient's forces in malignant dyfenteries, with what is found to do it by experience; and this quality belongs inconteftably to chicken-broth, though in other cafes it has manifeftly a con-trary effect. But if it be found neceffary to ftrengthen the patient more effectually, fome crumbs of bread are to be boiled in the broth,

and

and directly after eating it the patient fhould take every four hours a fpoon full of old white wine, which muſt not however be too ſtrong. The Frankfort, Mofelle, and Rheniſh wines are moſt proper for the Germans; the wine from the marquifite of Baden and the Lacote wine for us; and for the French, a wine very like the laſt mentioned, which comes from Grave in Guienne, a few leagues from Bourdeaux, on account of their refpective cordial, and at the fame time antifeptic powers.

Wine does as much good in this fpecies of dyfentery as harm in the others: though fome Swifs philofophers, quite ignorant of medicine, are as little informed of this, as of the flightnefs of fome and the malignity of other cafes in the dyfentery; and yet proud of their fophifms, think themfelves able to overturn with one whiff of their breath the pillars of phyfic, and with their fhallow doubts quibble the phyficians out of all pretenfions to the leaſt certainty in their profeffion.

In malignant fevers in general, according to the obfervations of her Britannic majefty's phyfician in ordinary, now Sir John Pringle, nothing exceeded the effects of wine in fuch patients,

246 A TREATISE on

tients, as were weak and deprived of all ftrength.
His patients fhewed a particular defire for fomething cordial ; when, together with a flow and
weak voice, they had had the fever a confiderable
time ; and nothing was then fo falutary as wine :
they did not long after any kind of food, but
yet eat with a great deal of pleafure a little foup
with crumbs of bread, provided there was wine
in it. On the other hand fuch as were delirious,
with a * quick voice, wild looks, fubfultus tendinum, or violent geftures, could not bear either
wine or heating medicines, nor even common
cordials. Now Dr. Pringle will have the malignant dyfentery treated in general like a malignant fever ; he therefore advifes the ufe of
wine in certain circumftances in this dyfentery :
he allows it in general in this diforder, when
the patient's ftrength is decreafing, and his
voice low and weak ; but he adds that we can
be never abfolutely fure of the effects of wine,
till we have tried it. Dr. Monro in like manner
made ufe of wine in the malignant fever with
great fuccefs, and in general the lateft and beft
phyficians in Great-Britain agree in this refpect.
In the malignant dyfentery one of the greateft
mafters of our profeffion, her imperial majefty's

* Thefe are Sir John Pringle's words : though the German (by miftake, as I imagine) has it " weak voice."

firft

firſt phyſician, baron Van Swieten, orders an ounce to be taken every hour of a diet drink, which conſiſts of half a pint of wine, a pint and a half of barley-water, one ounce of cinnamon-water, and ſix drachms of ſugar.

Acid liquors are uſually given in malignant fevers, as plentifully as ever the ſtomach and bowels will permit. But the lateſt diſcoveries with us have ſhewed, that acids given alone are noxious in theſe fevers, and more eſpecially in malignant dyſenteries; in which therefore the excellent Dr. Schintz in Zurich, ſeems with good reaſon to be afraid of fruit; though profeſſor Vater ſaw a malignant dyſentery cured with raw prunes. The bowels are in this dyſentery ſo much weakened by the venom which they harbour, that they can neither bear the ſame quantity of liquids, nor thoſe of ſo very emollient a nature, as in the other ſpecies of dyſentery. A too great quantity of liquor finds no paſſage out of the body, increaſes the anxiety, diſtends the belly, and cauſes a retention of urine. The ſame happens, when the liquor is only emollient; by which beſides the patient's weakneſs is increaſed. This ſame debility is alſo the cauſe, that the uſe of pure acids, which are otherwiſe an antidote to putridity, does more harm than good in the malignant dyſentery; for which reaſon the patient's

R 4 drink

should be neither in too great quantity, nor too emollient, nor too acid. A ptisan of fresh Seville oranges cut into thin slices, strewed over with some sugar, and infused in boiling water, possesses all the qualities here requisite; the peel is aromatic, the white has a kind of a corroborating bitterness, the juice is acid, and all this together has a very good effect. A great many succedaneums to this ptisan may be prepared by pouring water on a bitter substance, and then making it a little sour. But when the patient's loss of strength is extremely great, wine is the only acid that should be made use of.

Clysters of a laxative, emollient, fat, oily nature, are pernicious. A great many clysters should never be given, and those not large, at most not containing above seven or eight ounces: the only proper ones are such as consist of nothing but infusions of bitter herbs and flowers, as chamomile flowers, melilot, and burnet.

Some physicians, who first remarked that the watery pustles appearing on the skin were of service, and yet did not know, that instances of cutaneous disorders, that went and came according to the increase or decrease of the diarrhœa, are to be found in the writings of Hippocrates, and that Themison has long since advised the use of

cupping-

cupping-glaffes, ordered their patients to be cupped with the beft fuccefs. Others ordered their patients arms, thighs, and legs to be burnt with red-hot irons, as directed by Hippocrates, and very much practifed in the fixteenth and feventeenth centuries; which, in a very bad epidemy, accompanied with black fpots all over the body, that happened in England in the year 1513, was the only means of recovery. In our times we know the art of doing as much or more fervice with infinitely lefs pain. Galen advifes in the dyfentery every thing in general, that draws the morbific matter to the fkin, and many have followed this doctrine. Reftaurand publifhed 90 years ago divers obfervations on obftinate diarrhœas and dyfenteries, that he had cured not only with red-hot irons, but alfo with blifters; and Theophilus Bonnet fays, that thefe laft are the moft effectual of all remedies to procure an afflux of the humours towards the fkin.

However, for as much as I know, thefe hints have not been purfued; for Dr. Pringle and Monro made ufe of blifters merely for the pain. The honour of firft having revived the ufe of veficatories in malignant dyfenteries, belongs to two phyficians illuftrious for their inventions in the practice of phyfic, Dr. Hirzel of Zurich and Dr. Tiffot, though neither of them knew any

thing

thing of the other's difcoveries. Dr. Hirzel began with a woman, that in the malignant dyfentery had convulfions and fainting fits every
quarter of an hour, and during the intervals lay
in a perpetual delirium ; he delivered her from
this dreadful diforder chiefly by means of blifters;
and thus effected a cure, which to me indeed
appears confiderable, but the luftre of which
the doctor obfcures by making as dextrous and
important cures almoft every day. Dr. Tiffot
faw in many cafes his patients ftools and anxiety
diminifh, and their ftrength increafe, as faft as
thefe plaifters operated ; he therefore never
neglects this auxiliary in malignant dyfenteries,
except a great deal of pure and diffolved blood
comes away with the ftools.

All thefe remedies are fometimes not fufficient, when the patient's pulfe finks, his ftrength
is brought down, and he himfelf oppreffed with
anxiety ; the diforder then requires all the fame
remedies that are neceffary in malignant fevers.
The bark claims a place here above all the reft.

It is well known, with what fuccefs her imperial majefty's phyfician in ordinary, Dr. De
Haen, made ufe of this fimple in malignant
fevers ; and what great merit he has, in having
exactly determined the method requifite to be
followed in this cafe. Dr. Monro imitated him

in

in the Englifh army in Germany, and gave in malignant fevers in general, the bark in large dofes: he treated above 150 Englifh foldiers in this manner, and though he did not obtain his end with all of them, yet he found this medicine better than any other.

Dr. Medicus of Manheim, now refiding in Paris, a firft-rate medical genius of 29 years practice, has confirmed this doctrine to me by word of mouth (while this work was at the prefs) from his own numerous obfervations. I fhall find another place for my own obfervations; which tend to prove in like manner the excellency of the bark in malignant fevers, when the putrid matter is previoufly evacuated.

Degner, as eftimable as he is in many other re-fpects, in the fecond edition of his Treatife on the Dyfentery, judged in a very unphilofophical and indeterminate manner, on the ufe of the cortex in the dyfentery in general; as he only fupports his opinion with faying, that certain furgeons had killed with it a good many foldiers, that lay ill of the dyfentery. It is indifputable, that all furgeons do not underftand the catechifm of experience; but this I know, that great phyficians have made ufe of the bark in the ma-lignant dyfentery with good fuccefs. Dr. Monro,

as

as foon as petechiæ appeared, or the fever was obferved in any wife to remit, gave every four or fix hours a drachm of an electuary, confifting of equal parts of the ufual electuarium è cortice Peruviano, and diafcordium; or half a drachm of powder of bark, or 20 grains of its extract in the fpiritus mindereri, with four or five drops of the tinctura thebaica; and in the evening an opiate proportionate to the effect of the preceding dofe, and to the number of the patient's ftools. It muft be confeffed, that my friend Monro was not fuccefsful with all his patients; however, he found this method better than any other he tried. Dr. Tiffot gives in malignant dyfenteries the extract of bark diffolved in orange flower-water, but never in very large dofes, and not above two drachms in 24 hours.

The cortex is alfo principally of importance, when any external part of the body is feized with a gangrene. This does not feldom happen in malignant dyfenteries; and Dr. Baldinger, who has feen fo many cafes in the Pruffian army during the laft war, and thofe, contrary to cuftom, with fo great accuracy, has often remarked in fevers and diarrhœas, that the gangrene began at the tip of the nofe; that then the whole cartilaginous part of the nofe became of a dark-red colour like a cherry; whence it proceeded

proceeded to the eyes, attacked the cheeks, and
for the moft part killed the patient within five
or fix hours. I remember to have read a remarkable cafe that happened in England, and
is written in the Englifh language, which deferves to be repeated here. A brifk and tolerably healthy widow of a middle age, that had
lain a fortnight or three weeks very ill of the
dyfentery, and had not been fufficiently purged,
was attacked with violent diftractile pains,
deeply fituated in her legs and feet, and particularly in one leg and foot, which her phyfician
at the fame time found cold and ftiff. He gave
her directly internally the compound decoction
of fnake-root, according to the Edinburgh Difpenfatory, and ftrong aromatic fomentations for
her leg and foot ; notwithftanding which, the
next day the firft joint of all the toes belonging
to that foot was mortified ; the gangrene proceeded too round about the margin of her foot
under the little toe, and a broad fpot of a
brownifh-yellow hue appeared on the top of her
foot towards her great toe : her dyfentery was
ftill very violent. The decoction of bark was
given her immediately upon that, which fhe
took for a long time. This put a ftop to the
progrefs of the gangrene : a flight inflammation
appeared round about the margin of the gangrened part ; the dark-yellow fpot on the top
of

of her foot grew at firſt of a bright-red, and
aſſumed by degrees the natural colour of the
ſkin.   Her gripes and teneſmus decreaſed by
degrees ; her ſtools grew natural without the
aſſiſtance of any other purgative or anti-dyſen-
teric medicine than rhubarb, which was mixed
with the decoction of bark ; the mortified
parts ſeparated every where from the bone, and
the ſick perſon recovered.

Camphor deſerves a place next the cortex, or
in the ſame rank with it for raiſing the patient's
forces in the malignant dyſentery.   This is in
like manner a ſtrong antiſeptic ; and, according to Dr. Baldinger's frequent obſervations
made in the Pruſſian army, increaſes the effi-
cacy of the bark, and heightens its antiſeptic
power.   Camphor is very conveniently joined
with the extract of the bark, and even with
ipecacuanha; theſe three medicines may be given
at the ſame time, and even mixed together in a
mixture or bolus; or the two firſt be adminiſtered
likewiſe after the ipecacuanha ; which laſt is
particularly uſeful when the patient abounds
with mucus ; but the uſe of it may be diſpenſed
with, when no more mucus appears, and the
abdomen remains ſoft : but the camphor muſt
not be given in large doſes any more than the
extract of bark, and not above the quantity of
ſixteen

fixteen grains in twenty-four hours. Sometimes as an external application, with the fame view and with good fuccefs, we make ufe of a piece of flannel dipped in a bitter decoction boiled up with the theriaca, and applied warm to the fto-mach and abdomen; or even of a plaifter pre-pared with the theriaca alone.

When in the malignant fever in general, even though the patient at the fame time took the cortex and wine, the pulfe fank, and a de-lirium came on with other bad fymptoms, Dr. Monro laid the bark afide, and gave a cordial mixture with fifteen grains of mufk, ordering the wine to be boiled up with cinnamon. The next day his patients were better, their fkin was moiftened, their pulfe rofe, the feverifh fymp-toms went off by degrees with the ufe of the fame medicines, and they got well : the con-fectio cardiaca, fnake-root, with other medi-cines of that fort, had the like effect. I relate this however only with a view to mark out with one fingle ftroke of my pen the perfectly pecu-liar nature of a malignant fever for the ufe of fuch, as make a hotch-potch of all fevers, and then attack it with the like hotch-potch of me-dicines. It is well known how advantageoufly Bontius has fpoken of the extract of faffron in malignant dyfenteries, and how famous this ex-tract

tract is at prefent, fince Dr. Pringle's lateft obfervations, as much on account of its antifeptic
as of its cordial properties: but I tremble,
when I think, what havock phyficians, that
underftand nothing but receipts, will make
among their patients with all thefe medicines,
in diforders which they do not know, and in
cafes which they cannot diftinguifh.

Aftringent and conftipating medicines are of
ufe, in certain obvious circumftances, as well
in the malignant dyfentery, as in malignant
fevers in general.   Many perfons afflicted with
the malignant fever, have fometimes a diarrhœa
withal, which feldom terminates favourably;
and fome are even at that time feized with the
dyfentery.   A flight diarrhœa, that does not
very much weaken the patient, is indeed in general ferviceable, particularly if it appears at the
height or towards the end of the diforder.   But
one that is very confiderable, or actually de-
generates into a dyfentery, is extremely dangerous ; as every thing that retains the ftools increafes the fever; and as, on the other hand,
the continuance of the purging brings the patient very low, and indeed in a fhort time to the
grave.   In this cafe Dr. Monro was abfolutely
obliged to give an opiate after each purge.   Dr.
Pringle made fuch a point of gradually ftopping
by

by thefe means the purging, which appears towards the end of a malignant fever, that he added
a few drops of laudanum, or a fmall quantity of
theriaca to his alexiterial julep, or elfe exhibited a
fpoonful or two of a reftringent mixture. For
though this loofenefs might be looked upon as
ferviceable, it was ftill to be put a ftop to; if
the patient was too weak to undergo fuch an
evacuation. Dr. Pringle has alfo very often
obferved, that when it was ftopped in the above-
mentioned manner, the fick perfon fell foon
afterwards into a moderate fweat, that carried
off the diforder. In the worft cafes of the malignant fever, and efpecially when it is accompanied with the dyfentery, the ftools are often
bloody; in which dangerous fituation the Doctor
orders thefe fame medicines to be tried, if any
thing can yet be done for the fervice of the patient.
After frequent relapfes of the malignant fever the
blood was fo diffolved, that the patients bled violently at the nofe, and blood even came away with
their ftools : if they had a diarrhœa with it, Dr.
Monro joined diafcordium with the bark, giving
a dofe of opium at night, and at the fame time
neverthelefs exhibited the tincture of rhubarb.

Now it is really dangerous in this dyfentery
to proceed immediately without the preliminaries

S   above-

above-mentioned to reftringents. To be fure
the ftools have been fometimes retained by thefe
latter without the former having been premifed;
but they have alfo occafioned the greateft anxieties and a violent perturbation of the fenfes, and
have finally been the patient's death, or threw
him into a quartan ague, hectic fever, or oe-
dematous complaints. But when the violence
of the diforder was allayed, when the head-ach,
fever, tenefmus, convulfions, and the other
fymptoms remitted, and nothing remained but
an obftinate purging, in that cafe Dr. Vater
gave in the epidemy in Saxony I have fo often
made mention of in this treatife, the ipecacuanha
as an emetic with the greateft fuccefs; for it
often put a ftop to the diarrhœa : if it did not,
he gave with a very good effect aftringent powders compofed of radix tormentillæ, nutmegs,
and terra Japonica, in a mixture of theriaca, diafcordium, and cold water, (which he found
much more ferviceable than warm) and with this
cured his patients in a fhort time. Sometimes
however one is forced to add reftringents to the
corroborant remedies early in the diforder. In
very fevere cafes, when the patient's mouth and
œfophagus were threatened with aphthæ, and
when they were actually covered with them, the
ingenious Dr. Whytt of Edinburgh prefcribed

with

with fuccefs the confectio Japonica according to the Edinburgh Difpenfatory, with a ftrong decoction of the bark. And in general an extraordinary quantity of ftools in this dyfentery require narcotics and reftringents; on which account alfo baron Van Swieten prefcribes a grain of opium morning and evening.

But we muft here proceed with the greateft caution. Malignant diforders do not only appear at firft not fo dangerous as they really are, but all the remedies made ufe of with fuccefs by very great phyficians in thefe diforders in general, and efpecially in the malignant dyfentery, are a deadly poifon in ignorant hands, if its various fpecies are not diftinguifhed with the greateft penetration; or if the phyfician be ignorant that in the fame fpecies the fame remedies either kill or cure at different junctures. Millions of people might have been faved from an untimely grave; if in confequence of a more accurate knowledge of the true caufes of weaknefs in fevers, this great maxim had been underftood, that the patient is moft ftrengthened when the caufe of the diftemper is diminifhed; and that the reftoration of the forces is the firft, chief, and only view in all diforders with blockheads alone.

The

The catalogue of the medicines that have been recommended to this time downwards in malignant dyfenteries, is of an infinite extent. If we erafe all thofe which experience has fhown to be detrimental, we fhall fee, that all the reft agree in properties with thofe I here advife. I have only recommended the moft efficacious, and I believe they may very well be fubftituted for the reft, and do every thing that is expected from our art in malignant dyfenteries; for in this fpecies nature does nothing; and the reader has fufficiently feen, that fuch as defpifed all phyfic and phyficians, fuffered and died in the moft miferable manner. But it is upon this very account, that our art often falls fhort in this dreadful diftemper, as not being fupported by nature, efpecially when the phyfician does not attack the diforder immediately at its firft rife, and before it has irrevocably damaged the primæ viæ. Great phyficians have fully experienced, that there are degrees of malignity in which all methods and medicines are of no avail, and in which there are unexpected metaftafes of the malignant matter from parts, on which they had not fo pernicious an effect, to the brain, where they induce fudden death, at a time when the patient thinks himfelf almoft recovered. They honeftly own likewife, that they never durft make any certain prognoftic of the event in malignant

epidemic

epidemic dyfenteries; as on the one hand patients, whom they looked upon in a very bad light, and perfectly defpaired of, happily recovered; and on the other hand, fome died fuddenly, that had but flight fymptoms of the diftemper, and whofe recovery appeared to be certain.

The indeterminate idea commonly annexed to malignant diforders in general, and to the fymptoms of malignity in particular, is as great an evil, as the juft now remarked uncertainty in the thing itfelf. Throughout all Switzerland the common-rate phyficians call every diforder malignant, which they do not underftand; and according to them, every diftemper, of which their patients have died muft be malignant beyond all contradiction. Formerly all diftempers attended with puftules or fpots on the fkin, were looked upon as malignant without exception; and on this account they always endeavoured to cure them by fweating the patient with heating remedies. The fmall-pox and all fevers attended with any notable degree of inflammation, which were only to be cured by cooling remedies, were nick-named malignant, attacked with the moft heating medicines, and confequently, as Syden-ham faid, this fame word malignity did more hurt to the world than the invention of gun-powder. In order to expel this malignity, it

was

was formerly the cuftom at Vienna, (which has now in thefe later times taken a more philofo-phical turn, and that in my opinion chiefly by the means of fome philofophichal phyficians that practife there) to call the fratres mifericordiæ at the firft fight of the exanthemata, who, by the means of blankets, wrappers, &c. faftened the patients down to their beds fo effectually, and covered them up fo clofe, that they could neither ftir nor breathe; fo they had the exan-themata very happily out upon them, and died. But I alfo often find here and there in modern writers fuch confufed indeterminate ideas of ma-lignity, and thofe fo very hypothetical withal; that I fhould be aftonifhed at the want of a philofo-phic turn of thought in this our moft philofo-phic age, on this and other occafions, if I did not fee with my own eyes how many phyficians of the higheft reputation know as yet nothing of the improvements made in our days.

In fine, the abufe of many medicines, which in fome cafes are of real fervice, and in a thou-fand others extremely deftructive, ferves to prove here in its right place the following maxim, that a partial notion of phyfic is juft as dangerous as one that is entirely falfe. I have, from an experience of their virtues, recommended cor-roborant and ftrengthening remedies in the cure

of

of malignant fevers in general, and of the malignant dyfentery in particular. But I am afraid
of their being abufed, a fate that has often been
known to attend them ; as the general rules for
the cure of malignant fevers are extremely limited; as a perturbation of the fenfes may proceed from two quite oppofite defects, the one
from large and repeated venæfections, the other
from heating and cordial medicines having been
given too foon ; and likewife, as the imprudent
and on that account very common ufe of wine,
as furely hurts and kills the patient in a malignant dyfentery, as in an inflammatory fever.
Phyficians, that want for genius and experience,
can never perceive the precife juncture of time,
in which a rapid and very dangerous diforder
requires the ufe of wine, in which there is that
peculiar fpecies of weaknefs, that only yields
to the ufe of warm and cordial medicines.
Nothing is eafier in this cafe than to commit
an error, and the certain confequence of the
leaft error of this kind is death.

I now turn to the cure of that period or fpecies of the dyfentery, which is called the chronic, and in which as many blunders are committed as in any other ftate of this diforder.

S 4 It

It is infinitely difficult to cure people of the dyfentery, that have been ruined by a falfe method of cure with carminatives, cordials, aftringents, and narcotics, and thence are afflicted with flight inflammations in the bowels, or a kind of paralyfis in them, have indeed little pain, but yet are daily and frequently harraffed with painful ftools, and whofe forces are vaftly exhaufted. If the phyfician is called too late, or the patient neglected, or badly treated; if he has, from a too great lofs of his fluids, a flow weak pulfe, vaft general debility, a rough dry cruft on the tongue and the infide of the mouth, excrements, in which the villous coat of the inteftines may be diftinguifhed, and a great relaxation of the bowels, he is then without doubt in much danger of his life; and in that cafe likewife, in the opinion of many great phyficians, there is nothing more to be done, than what is ufed to be effayed in a fuppuration of the inteftines : as in this high and dangerous degree of the dyfentery vomits and purges are very feldom of any ufe; and as too at that time opiates have hardly any effect either in relieving the pain, or putting a ftop to the flux. The reft of the cure fhould be left to the fick man's natural forces, of which he is perhaps not yet entirely deftitute; with which however the quite exhaufted,

exhaufted, but patient fufferer is fometimes ftill fupported for many weeks, and even months, on the brink of the grave, and by degrees brought to a perfect recovery. Dr. Monro fays, that he was never more fuccefsful with any diforder in the Englifh army in Germany than frefh contracted dyfenteries; but when they had continued fome weeks, and were in a manner become chronical, all his endeavours were then frequently fruftrated, and a great number of his patients died. Dr. Cleghorn too at Minorca found all dyfenteries, that were not cured fpeedily and in the beginning, at leaft obftinate, and in fpite of a great number of highly extolled fpecifics, too often fatal. The Englifh phyficians and furgeons, that were laft war in the American fervice affured Dr. Monro, that they had full as bad fuccefs with dyfenteries of long ftanding in America, as he in Germany. However it muft not furely be concluded from all this, that every chronic dyfentery is a loft cafe, and on that account fhould be given over, and the cure not attempted; for many have with great care, and by the ftrength of their conftitutions, got over even this diforder by degrees, and enjoyed their health again; efpecially thofe that held out the winter, and remained alive at the appearance of the warm weather.

The

The indication towards the cure of the chronic dyſentery is in general, to evacuate the putrid humours, and at the ſame time ſtrengthen the bowels. In ſuppurations of the inteſtines eſpecially, we ſhould endeavour to cleanſe, and withal heal up the ulcers. But all this is not ſo eaſy to be done : for a great many attempts have been made to cure this dyſentery, many have failed, and ſome were not uſeleſs. I ſhall paſs over ſuch as have miſcarried, and ſhall recommend thoſe that ſeem to promiſe ſomething more; after that I ſhall give the moſt univerſal and beſt method of curing this ſpecies, and laſtly add a few more cautions on this head.

In difficult caſes of the chronic dyſentery Baglivi adviſes to pour turpentine on live coals, and to take in the vapour of it at the poſteriors, and hence he promiſes a certain cure; to which however I do not give a great deal of credit. Huxham, with many others, at firſt makes uſe of warm water, as it cleanſes the inteſtines very well, and likewiſe paſſes very eaſily into the blood; but when once the acrid humours are evacuated, he then very rightly adviſes cold water, and aſſures us, that with this alone and opium he has ſometimes, after the proper evacuations had been premiſed, compleated the cure. This method ſhould by all means be

tried

tried in the chronic dyfentery; and while this
work is at the prefs a very remarkable inftance
has happened to prove the falutary effects of
cold water in obftinate dyfenteries. Dr. Smith
of Bellikon, phyfician in ordinary to the princely
foundation at Einfidlen, one of the moft learned
and judicious phyficians in Switzerland, writes
me word, that during the epidemy in 1766, he
had treated with the greateft care a woman of
63 years, that lay ill of the dyfentery, with various medicines during the fpace of ten or twelve
days; but when at length he faw the diforder
was not in the leaft diminifhed, he ordered her
to drink a glafs of quite cold water every four
hours, and allowed her no other nourifhment
than lukewarm milk. She did this three or
four days with fuch fuccefs, that her ftools grew
much lefs frequent, and there was no more
blood to be feen in them; her pain and bearing
down diminifhed, and at length the patient was
compleatly cured by this method, which, by its
noble fimplicity, does honour not only to the
phyfician, but likewife to the art of phyfic
itfelf.

The fimaruba has been moft made trial of.
Juffieu and others make a great noife with this
bark; perfons that have been plagued with
diarrhœas, and chronic dyfenteries for many
months

months and even years together, Juffieu has for
the moft part 'fet up again more fpeedily and
certainly with this, than with any other medicine; and that without occafioning any indifpofition, without the leaft hindrance of any of
the natural functions, and without the leaft bad
confequence enfuing ; he even cured diarrhœas
with this bark without the leaft prejudice in the
midft of the hemorrhoidal flux, or the men-
ftrual evacuation ; and he actually gives it out
univerfally, that this bark is a certain cure for
inveterate, watery, mucous diarrhœas, proceeding from a continual convulfive motion of
the inteftines, without prejudice to the ftomach,
and without the leaft irritation in the bowels.
Du Buiffon has long made ufe of it in all im-
moderate alvine fluxes, in old loofeneffes with
indigeftions, and in general in all diarrhœas of
long ftanding with a good effect.   Dr. Winter,
formerly phyfician to the prince of Orange at
the Hague, and profeffor in Leyden, cured
three perfons with the fimaruba within a few
days of an extremely obftinate, though mild
dyfentery, in which ipecacuanha and rhubarb,
paregoric, aftringent, and other remedies had
not been able to do any thing in many months.

But all this requires fome limitations.   The
fimaruba does not always effect what is expected
from

from it ; for during the impreſſion of this work, my advice has been ſent for from Germany in a tedious dyſentery-like diarrhœa, attended with manifold bad ſymptoms, with which a gentleman was ſeized, who had from his youth upwards been extremely hypochondriacal and weak, and previouſly to this had had a continual diarrhœa in 1763 and 1764, and in the year 1765 a violent dyſentery, after which attack the ſimeruba ſeemed rather to increaſe than diminiſh the malady. In my judgment this medicine does beſt in cafes, where the patient requires merely to be ſtrengthened, and not when a detergent is wanted, for in that point it is excelled by the tincture of rhubarb. In every diarrhœa and dyſentery, where the bowels harbour a corrupt putrid matter, the ſimaruba is either uſeleſs or pernicious ; but when this is cleared away, it braces the relaxed bowels, and all their veſſels. It is extremely hurtful in dyſenteries, that are attended with ulcers in the bowels. It has likewiſe been obſerved, that the ſimaruba operates much better and is more effectual in ſuch chronic dyſenteries, as are at the ſame time bloody ; but when after the blood ceaſes to appear, the ſtools remain liquid and mucous ; if the caſcarilla be then added to the decoction of ſimaruba, the ſtools are then much more opportunely diminiſhed, and by their

united

united force the cure is alfo compleated much more fpeedily and certainly.

All things taken together, feleĉt, genuine, and undamaged fimaruba is no bad remedy with the above-mentioned provifo in dyfenteries of long ftanding. The beft method to give this bark to the fick is this ; take two drachms of it, and let them digeft two hours in a pint of water in a warm place; then boil the whole for half an hour, pour it through a fieve, and give one half of it in the morning and the other in the evening ; continue every day in this manner, and in cafe it fhould be requifite, for the fpace of three or four weeks; if it be obferved, that with the ufe of this decoĉtion the urine paffes more plentifully, and grows of a paler colour, it may be looked on as a fign, that the remedy takes effeĉt, and that the loofenefs will foon ceafe. Others mix two drachms of this bark cut very fmall with two pints of water, boil it to one third, and order it to be taken warm at four times in the fpace of a day ; or elfe they exhibit half a drachm of the powder with two ounces of water or fyrup of maidenhair, and repeat the dofe till the cure be compleated.

When Degner had patients, to whom every thing, and even the fimaruba itfelf had been
given

given in vain, or without the wifhed for effect,
and whofe bowels were grown exceffively weak
and relaxed, he then made ufe of corroborating
remedies, and even of fuch as were abfolutely
of a reftringent nature, as cafcarilla ·and terra
Japonica with great benefit. The cafcarilla is a
good ftrengthener, though in other refpects
Stahl's followers in Germany have made too
much of it. The terra Japonica requires more
caution, being of an aftringent nature; but it
is not at all to be rejected in cafes, where there
is need of aftringents. Extract of Campeachy
wood diffolved in mint-water, and lime-water
diluted with milk, were very effectual too in
this refpect.

In dyfenteries attended with exulcerations,
trials have likewife been made, which I cannot
pafs by in filence. In that degree of the dyfentery, in which the body was exceffively weakened
and exhaufted of all its ftrength, in which the
pieces that came away from the tunica villofa of
the inteftines, betrayed a very confiderable relaxation of the bowels, in which, inftead of
blood, a thin purulent matter was feen in the
ftools, Degner found no purge of any fervice
befides manna and the extract of rhubarb.
With thefe he gave daily vulnerary herbs in-
fufed in water; alfo in the interim extract of

bark

bark and cafcarilla; and in this manner, in the
fpace of fix weeks or two months, he conceived
hopes of fome of his patients, that they might
ftill get perfectly well. Gum arabic was of good
effect in the chronic fuppurative dyfentery, and is
therefore with much reafon greatly prized in
fuch cafes, and given with the ufual decoction,
or in barley-water: Dr. Baldinger, from whofe
treatife, known to me but too late, I have learned
fo much with the greateft gratitude, found this
gum very good, when the bowels were eroded.
Gum tragacanth was in like manner falutary
in thefe cafes. Maftich was a good and fure
remedy, partly as a corroborant, and partly as
a balfamic; Dr. Baldinger has alfo obferved,
that the balfam of maftich (as it is called) according to the prefcription of his Pruffian majefty's phyfician in ordinary Cothenius, is a very
efficacious remedy if it be made ufe of as an
aftringent, with prudence. In abfceffes of the
ftomach, that were formed in confequence of
the dyfentery, and were burft, Mead got his
patients well with the balfamum locatelli.

The method of cure found from mature experience, to be moft univerfal and beft in chronic dyfenteries, remains ftill to be taken into
confideration. Among the remedies that come
under this predicament, purgatives are very neceffary

.ceffary even during the ufe of quite different ⎱ /
remedies, or elfe given at intervals from time to ⎰
time. Net only the putrid humours, but likewife the hard excrements, that often remain in
the cavity of the inteftines, muft be evacuated ;
if this be neglected to be done, the patients are
often feized with a ficknefs at ftomach, tormina,
and a violent loofenefs. If they fhould at any
time have a pain in the bowels, and little mor-
fels of hard excrement come away from them,
this is moft commonly a certain fign, that evacuations are neceffary; and for the moft part
they give relief to the patient. Sicknefs at ftomach, with the other circumftances, that ufually
attend it, requires an emetic to be given, before,
the patient be purged; in cafes where the patient is very weak, or in violent pains and tenefmus, clyfters are made ufe of. It has been
remarked, that in complaints of long ftanding, and already arrived to a high pitch, the
patient has appeared to get better; but on evacuation of the above-mentioned globular hard
excrements, has had a relapfe; as thefe fcybala
came away for many days together, and caufed
thereby a perpetual irritation : they fhould therefore be expelled in proper time either by a good
dofe of rhubarb and manna, or elfe with tamarinds, or the frequent exhibition of oily clyfters.

T   Some

Some gentle purge ſhould always be given al-
ternately with the other medicines, when there
is no ulcer, nor any thing worſe in the bowels;
and this ſhould be continued, till the belly-ach
and teneſmus is entirely put a ſtop to. I do
not know from my own experience, what power
Glauber's purgans catholicum, ſo much eſteemed
by Dr. Werlhoff, in the doſe of half a grain or
a grain, in diarrhœas of long ſtanding, may
have in this reſpect; but this I know, that the
tincture of rhubarb made with water is with
this view a very excellent medicine. I have
often ſeen, that it braces the patient more, and
even cures the purging more certainly in this
ſtate of the diſorder, than aſtringents.

Brockleſby made uſe of a method in the chronic dyſentery, which deſerves notice, as it conſiſts in the junction of two medicines, which are
ſeldom ſeen together. Every morning and evening he gave two grains of opium made into pills
with three grains of powder of ipecacuanha,
and found this remedy uncommonly ſerviceable
to many people. The ipecacuanha given in
this manner became a mere gentle purge;
while on the other hand the opium alleviated
the irritation occaſioned by the ipecacuanha and
morbific matter. The Doctor found no remedy
more

more extenſively uſeful in the chronic dyſenteries, that came under his obſervations, when
the purging and bloody ſtools were ſtill conſiderable, but the fever was quite gone; and he
affirms, that no one can conceive without making trial of it himſelf, how much the mitigating
power of the opium corrects the irritation of the
ipecacuanha, and on the other hand is itſelf corrected by it. I know from my own experience
that thus far is true; but this medicine has
often proved unſerviceable.

In general the following method ſeems the
moſt likely to cure dyſenteries of this ſort, when
they are not arrived to too high a pitch.

The patients muſt be content with a thin low
diet, conſiſting chiefly of milk, rice, ſago, —
and ſalep. They may be indulged in weak
meat-broths, and chicken or veal, when they
find themſelves actually in a ſtate of recovery.
Their uſual drink muſt be barley-water, thin
rice-gruel, toaſt and water, or almond-emulſion. They muſt go warm cloathed, and be
particularly careful not to give occaſion to the
repulſion of the perſpirable matter; for errors
in diet of every ſpecies and catching cold are the
moſt uſual cauſes of relapſes.

Gentle

Gentle purges fhould be given from time to time either of manna or falts, or elfe of manna diffolved in almond-emulfion; or rather tincture of rhubarb; and fometimes a gentle emetic.

Among the ftrengthening and gently aftringent remedies the bark combined with aftringents and opium is of fervice to fome perfons; to others aftringent and anodyne clyfters; others do better with other things; and many find themfelves better, when they do not take any medicines of this fort at all.

The patient fhould occafionally take opiates, go into the frefh air, and ride with moderation by way of ftrengthening the bowels.

Dr. Monro, the inventor of this method, has feen cafes in this dyfentery (at the commencement of which the patient had gone through a due courfe of evacuation) cured with nothing elfe than broths, white meats, riding out every day, and a glafs of good wine. But he remarks very earneftly withal, that this method was ferviceable only in flight cafes, which had been previoufly palliated by the means of evacuations.

Dr. Brocklefby indeed is more favourable to wine in chronic dyfenteries, than my friend Monro.

Monro. In all thofe dyfenteries of long continuance, where the patient is quite worn out
and wafted away, which followed thofe bilious
fevers that were in particular very frequent in
the year 1758 after the return of the Englifh
from the coaft of France, Dr. Brocklefby found
port wine mixed with water very neceffary; and
often allowed his patients a pint and a half of it
mixed with a fufficient quantity of water every
24 hours during the fpace of three weeks or a
month. This with a pleafant decoction of cinnamon, orange-peel, and other aromatics
boiled in fpring-water, and given in a proper
dofe, was an excellent remedy, but was only
prefcribed to fuch as had no fever in the leaft.
The foldiers took fometimes the fpecies aroma-
ticæ in the dofe of ten or fifteen grains every eight
hours in this grateful decoction, with a view to
warm their cold and relaxed bowels, to put their
blood in motion again, and reftore to the folids
their wonted force. But when the purging continued, and the tenefmus was joined with it, (a circumftance, which at this time was not at all unufual) the Doctor found it indifpenfibly neceffary to
prefcribe again the gentle purges of falts, manna,
and fweet oil; and to repeat them in proportion to the patient's forces, till fuch time as the
tenefmus was over; which moftly happened in
a fhort time after. Yet having opened two

T 3   perfons,

perfons, that died in this juncture, he found in
both of them the rectum inflamed in the higheft
degree fome inches above the anus, though their
fever had ceafed a long time ago. A new proof,
how cautioufly we fhould proceed with wine
even in thefe old and tedious cafes of the dyfentery.

But unfkilful imitators fhould not only be
cautioned on the fubject of wine, but alfo con-
cerning the abufe of aftringents, and that even
with refpect to the chronic dyfenteries here
treated of.

One cannot be too cautious with reftringents.
Some years ago profeffor Schobinger had a young
lady of quality at St. Gallen for his patient, in
a flight but tedious dyfentery ; after copious
evacuations this worthy phyfician, fo little known
in his own city, at length gave her corroborants and gentle aftringents ; the Peruvian bark
mixed with a certain quantity of cafcarilla, the
fpecies hyacinthi, and the bolus armena, all in
very moderate dofes, and at a time, when the
purging and pains were almoft quite gone;
notwithftanding which there enfued on the ufe
of thefe medicines a flying gout, that lafted
three weeks. Brocklefby confeffes, it has but
too often happened to him, notwithftanding the

caution

caution with which he made ufe of aftringent remedies, that inftead of fhortening the diforder he only prolonged it, brought the fever on again, and was obliged to begin afrefh with vomits and purges. The mifcarriage of Dr. Monro's various endeavours to cure the chronic dyfentery appears too moftly to proceed from the ufe of aftringent or conftipating medicines. Even in that uncommonly mild chronic dyfen-tery at Java, defcribed by Dr. Laurich, which in gentlenefs much exceeds moft of our European dyfenteries, reftringents are extremely noxious. The phyficians of the country, as well as the European phyficians fettled there, have recourfe to them in this dyfentery. The Indian phyficians make ufe of the fruits called billingbing, macandou, nimbo, carambolas, and jangomas ; with thefe and other medicines of the fame kind they put a ftop to the loofenefs, without having prefcribed purgatives before-hand, and that to the great prejudice of the fick. Even the European practitioners there, who are moftly furgeons in the fervice of the Dutch Eaft-India company, fall into the like errors ; and a manual printed in the Dutch tongue at Middleburg for their inftruction, teaches them thefe very errors. Their moft efficacious remedies are the bolus armena, terra Japonica, a kind of terra figillata from Spain,

dragons

dragons blood, burnt harts-horn, corallium rubrum, the peels of unripe pomegranates, folid opium, and the infpiffated juice of floes : from all and each of which remedies Dr. Laurich faw an erofion of the bowels, a confequent fuppurative fever, the moft terrible fiftulæ in ano, and very often death itfelf take rife in this otherwife flight though tedious dyfentery. It is confequently a very neceffary and univerfal caution even in the chronic dyfentery, never to give aftringents without being firft perfectly convinced, that the peccant matter is evacuated, and that the fole caufe of the complaint is a laxity of the fibres.

I fhall now conclude this long difcourfe by farther taking into confideration fome new remedies, and laftly the various fpecifics recommended in the dyfentery.

The vitrum ceratum antimonii was firft made known as a powerful remedy in divers diftempers, and chiefly in the dyfentery, in confequence of the experiments of Dr. Young, Francis Pringle, Simpfon, Paifley, Stephen, and Gordon, which were inferted by Dr. Pringle in the Edinburgh Effays. Dr. Young takes an ounce of the glafs of antimony in powder and a drachm of white wax ; the wax being previoufly

melted

melted in an iron ladle, the powder is caſt into
it; the whole maſs ſet over a gentle fire without
flame for half an hour, ſtirred about with a
ſpatula continually, then taken off, poured on a
ſheet of white paper, and reduced to powder.
Of this powder Dr. Young gave in the dyſentery ten or twelve grains to adults; but uſually
for greater certainty began with ſix grains; to
children of ten years of age he gave from three
to four grains; and to thoſe of three or four
years of age from two to three. In general the
powder given in this manner occaſioned a ſickneſs at ſtomach and vomiting: moſt people
were purged by it; though ſometimes the cure
was compleated without any ſickneſs or evacuation. When it had operated too violently, the
Doctor omitted the uſe of it for a day: ſome
were cured with one ſingle doſe, others required
five or ſix; eſpecially when the firſt doſes were
too weak. He gave this medicine faſting, and
forbid all drink for three hours after it; but allowed warm water as with other emetics, when
the patient was ſick at ſtomach, or felt an inclination to vomit: the diet was the ſame, as
is uſually preſcribed in the dyſentery.

Dr. Francis Pringle's, Brown's, and Simpſon's experiments agreed with Dr. Young's,

and

and were extremely favourable to this remedy.
Dr. Simpſon was aware, that on account of the
variety in the ſpecies of the dyſentery, it could
not be ſerviceable to every body; and yet he
looked upon it to be in moſt caſes as great a
ſpecific in the dyſentery, as the bark in intermittent fevers, and external mortifications.
Dr. Paiſley at firſt made uſe of this remedy prepared in the ſame manner with much ſucceſs;
but afterwards, in purſuance of another receipt,
he only rubbed the ladle over with the white
wax, and did not reduce the glaſs to powder,
before he put it into the ladle; after having
held it the uſual time over the fire, the wax
ſtuck faſt to the ladle, and when all was cold,
he reduced the glaſs to a fine powder. He had
ſeldom occaſion for more than three grains of
this preparation, and never gave above five to
ſtrong perſons. In this way it operated full as
well; and, notwithſtanding the ſmallneſs of
the doſe, in the ſame manner as the other; and
Dr. Paiſley cured with it a great number of dyſenteries. Four or five doſes for the moſt part
perfected the cure, if they were uſed in time;
if the diſorder was of longer ſtanding, he was
forced to give ten or fifteen doſes, and never
ſaw any bad effect from it. Dr. Stephen, out
of a hundred and ninety perſons that he treated
with

with this remedy, loft but one. Dr. Gordon cured fome hundreds of people with this medicine in fmall dofes, and fince that it has never mifcarried with him; except in one or two cafes, in which he had proceeded rather too far with his patients; he commonly gave three grains of it, and never above five; a dofe or two was often fufficient, and he feldom gave three; he ordered it to be taken in the morning, and it was often two hours, before it operated; fome were purged by it, others both purged and vomited, and were fick for fix or eight hours together after it: at night he always gave a good dofe of opium.

Since that time this medicine has been tried in various manners over all Europe. La Mettrie, who is fo great an enemy to aftringents and opiates, is on the other hand very fond of emetics in the dyfentery; and efpecially thofe prepared with antimony but diffolved in a great deal of water, which rather act upwards than downwards, and which he looked upon as milder than rhubarb, as they rid the body at once of a fharp ftimulating matter. Even when a gangrene was aprehended, in obftinate dyfenteries La Mettrie made ufe of the vitrum ceratum antimonii in the quantity of a quarter of a grain, and vaftly extols its virtues even in

taking

taking away the pain; especially when the body has not been previously well purged either by nature or by art. Many experiments were made besides in France and Germany; and very lately an ingenious physician, Dr. Lentin, has found this remedy of great service in the dysentery; the Westphalian peasants, a people as tough as leather, can however bear but ten grains of it; and even our egregious Conrad Rahn honours it with his approbation. I have related in the sixth chapter the latest experiments made with it in the dysentery of 1765.

Now these experiments seem indeed to promise a great deal, but yet posterior experience has very much limited the reputation of this medicine. Her Britannic majesty's physician in ordinary, Dr. Pringle, whom we have to thank for the publication of these experiments in the Edinburgh Essays, found this medicine the most specific of all emetics with the English army, not only in giving relief to the stomach, but to the bowels too, if exhibited in the beginning of the disorder. But notwithstanding his being convinced, that it was a medicine of great power, he could not help being solicitous on account of its constant boisterous operation towards the end; and wished that he was able to cure his patients with more gentle remedies,

though

though they fhould chance to be flower in their operation. He therefore limited its ufe to ob-ftinate cafes; and faw that it did good fervice, where other things did none at all; if the bowels were but in a tolerable good condition, the patient but little feverifh, and not too weak. He makes withal the following good remark; that this remedy is attended with an inconvenience common to all preparations of antimony, namely the difficulty of affigning a proper dofe for it; as a middling dofe is at one time too fmall, and at another time too large. Dr. Eller made ufe of this medicine with two ftrong perfons ill of the dyfentery with the greateft fuccefs; on the other hand he found it very difficult with others of his patients to fix on a proper portion; as fometimes the fame dofe excited neither vomiting nor purging, and fometimes produced both but too violently. Dr. Monro found in the Englifh army during the laft war, that this remedy operated much too powerfully; and therefore it was almoft entirely laid afide.

Mr. Geoffroy at Paris has indeed endeavoured to find a correftor for this medicine, which, as it was imagined, would be able perhaps to prevent the indeterminate mode of its operation. Without doubt the exaft mixture of the wax with the vitrum antimonii, renders this other-

wife

wife very terrible medicine a falutary one, and
on this account Young's receipt is better than
Paifley's : on the other hand Geoffroy has fhewn
a ftill better method of involving this glafs by
the means of an oil, with which the glafs is to
be levigated on a porphyry-ftone.   But with all
this, though it muft be owned to be an excellent medicine in the bilious, and fometimes too
in the chronic dyfentery; it is, partly for the
reafons above-mentioned, and chiefly on account
of the caution with which it ought to be ufed in
inflammatory or even malignant dyfenteries, at
all times a dangerous remedy in ignorant hands,
and not always advifeable in the moft judicious.

The orchis, that comes to us from Perfia
under the name of falep, is alfo reckoned among
the new remedies for the dyfentery. Du Buiffon,
who got this root directly from Mocha, did
not look upon it in reality as a root, but as a
kind of. fig dried in the fun : my great mafter
baron Haller, takes it for what it really is, and
calls it the Perfian orchis.   It has the property
in common with our, and even the Swedifh
orchis, of being very clammy and full of a thick
mucilage ; the powder of it mixed with a good
deal of water, and fet over a very gentle fire,
turns to a jelly, and is looked upon to be full
as mucilaginous as tragacanth.   It is on that
account

account without doubt of ufe in the dyfentery,
whenever a medicine of this kind is neceffary;
but it poffeffes too at the fame time a gentle
aftringent nature, and therefore muft be exhibited with great caution. I know very well,
that after being ufed for fome time it binds the
body, and by caufing the pains to return makes
evacuants neceffary.

Sago, a feed that comes to us from Japan,
Ternate, and Amboyna, and which in the fame
manner turns to a jelly, is efteemed for the like
quality. It does not only affwage the pain, but
is likewife nourifhing; and mixed with fugar
and lemon-juice, a very palatable remedy. But
there is nothing extraordinary excellent in it;
and it is as little fpecific as the Perfian orchis.

The gitta gambir, which muft not be confounded with the gutta gamba, is in like manner a new remedy much efteemed in the dyfentery. It confifts of little lozenges prepared in
the ifland of Java, and which are fufpeéted to
be chiefly compofed of terra Japonica, or the
juice of the catechu-tree, or at leaft of fome
parts of that vegetable. The greateft partifans
of this remedy allow, that it does not cure the
dyfentery without other auxiliaries: it is befides
very dear, and muft be taken in greater quantities

tities than the terra Japonica, with which it chiefly agrees in properties, and on that account ought to be rejected in most cafes.

The bark of the tree mangoftan, which has been tranfplanted from the Molucca iflands to Java, and at Batavia is looked upon as the greateft ornament of their gardens, is likewife reckoned among the new remedies for the dyfentery. It has fome refemblance to pomegranate-peels, but is for that very reafon to be rejected in moft cafes.

The codaga pala, or the conefti bark, is confidered in Ceylon and Malabar as a powerful remedy in the dyfentery, and is very much efteemed in England, at leaft in diarrhœas. This medicine is bitter, and as fuch, may have its ufe in fome dyfenteries, when the primæ viæ are previoufly cleanfed ; but it has not been found more effectual than other bitter aromatics. It has fometimes even fhewn a torpefying power, and once in the prefence of Dr. Brocklefby, given in the quantity of two drachms, it caufed the fpafmus cynicus.

The aromatic bark of the guyava-tree is thought to be very ferviceable, efpecially in bloody dyfenteries : I was alfo told at Paris
about

about fifteen years ago, that they had begun to make ufe of an American root, called Poegereba, with the fame view. But who does not know, that many dyfenteries totally differing in fpecie from each other, are attended with blood?

But I here break off with pleafure; and rather defire my readers to ponder the remark fo often made, that a naufeous heap of noftrums rather ferves to render the phyfician doubtful and undetermined, than to anfwer the end propofed; and that a phyfician, who underftands his bufinefs, will obtain his purpofe infinitely better in all refpects with a few well-chofen remedies, than an ordinary practitioner with all the riff-raff of the apothecaries fhops.

I come at laft to the fpecifics, as they are called, the only thing in the whole art of medicine, on which the party that is fet up againft Dr. Tiffot, and which fo manifeftly endeavours to eftablifh empiricifm, at prefent chiefly founds its reputation; and which therefore in obftinate diforders looks for one fpecific after another with no other purpofe, than with hopes at laft to find the very medicine, that God has created folely for the ufe of the very patient, they chance at that time to attend.

U   I

I know from my own experience the truth of
what Dr. Tiffot fays, that there is no diforder
in which there are more pretended infallible fpecifics recommended, than in the dyfentery; that
there is no one that does not trumpet forth the
praifes of his own noftrum, prefer it before any
other, and promife with the moft perfect affurance
to cure a tedious diforder of long ftanding, of
which he has not the leaft idea, in a few hours,
with a remedy, the effects of which are entirely
unknown to him: while the poor patient, full
of fuffering and anxiety, and teized to death,
takes every thing from every body, and thus
out of fear, wearinefs, or complaifance, poifons
himfelf. But I know likewife from my own
experience, how dangerous it is to attack openly
the profeffors of an art, in order to fhew them
the prejudices that have infinuated themfelves
into it: for this is furely to attack the felf-love of
fuch, as entertain thefe prejudices; and thus by
forcing them to defend themfelves, we create
ourfelves as many enemies, as there are perfons,
for whofe information we wrote; and ftir up as
many antagonifts, that join in open cry againft
us, as there are people, that moft ftand in need
of our inftructions. The danger of fuch a procedure is emphatically fet forth by a Genoefe,
as famous for the acutenefs of his imagination
as the uprightnefs of his conduct, in a treatife
lately

lately publifhed, and very much wanted, on
" the neceffity and method of curing the art of
" phyfic of the very dangerous diforder of fraud."

I am indeed as far as any other phyfician of
probity from leffening the reputation of a medicine's virtues, if its effects do but conditionally
anfwer to the character given of it. But I hope
I may be allowed to doubt with difcretion, as
long as I am not convinced; to chufe out
of many remedies thofe few, that conduct me
in a fure manner to my propofed end; and to
be angry at the blindnefs of fuch, as in defperate cafes cry up the forrieft trafh for infallible
remedies, becaufe they confeffedly would not
be able to do any hurt in very flight cafes; or
that recommend a thing, which is indifputably
ferviceable in one fpecies of a diftemper, likewife in another fpecies of the fame diforder, in
which it is manifeftly fatal. In fine, I hope to
be permitted, in this labyrinth of human opi-
nions, never to fubject myfelf blindly to the fentiments of any other man, to look on the ftate
of the learned in general as a republic, and that
a free one ; and to proceed on the road to truth
with the greater caution, the eafier it is to ftep
afide from it into the path of empiricifm. Who
is ignorant, that the virtues of medicaments are
never abfolute and infallible, but merely relative,

and

and depending on the nature of the malady, and the patient's conftitution ; that on that ac-count there is fcarce any remedy to be found that is univerfal in all cafes, though fome are neverthelefs poffeffed of very extenfive powers ; and that in general we are not fo much in want of a ftore of medicines, as of fkill to make a pro-per choice of them.

The arguments, which phyficians have brought to defend their moft efteemed remedies, in the dyfentery, are perfectly ridiculous. Formerly antacids were efteemed in the dyfentery on the ftrength of that falfe maxim, that this diforder arofe from an acid ; for which reafon all four things were forbidden ; though nothing could be indicated more efficacious, with regard to its fo often putrid nature. The authors of the Breflau Journal of 1699, 1700, 1701, and 1702, fay very ferioufly : as the dyfenteric matter is acrid, fharp, four, and irritating ; it is very clear, that antacids fhould be ufed here, among which coral, the Silefian terra figillata, the prepared criftalla montana, periwinkles, and crabs-eyes, deferve the preference ; that is to fay, remedies, which for the moft part are endued with the property of increafing the pu-trid nature of the dyfentery. In like manner we fhall find here and there fome of our Swifs
                                        phyficians,

phyficians, who prefcribe nothing in our putrid fevers, as they are called, but thefe fame abforbents: in their opinion they fweeten the blood, which however, according to Dr. Pringle's experiments, they manifeftly render putrid, and in the mean while let the bilious matter, which ought to be evacuated, lie peaceably in the body. However the doctrine of acidity, that ancient gratuitous caufe of all diforders, was fcarcely banifhed; when the phyficians took it into their heads, that it was neceffary to make ufe of reftringents in the dyfentery. Accordingly they felected from other aftringents fuch as at leaft better agreed with this fyftem, coral, burnt harts-horn, and the like; but fad experience has fhewn us, that thefe medicines are not even reftringent. The terræ figillatæ indeed are, and by that means ftop the flux, caufe the heart-burn, vaft anxiety, and often death itfelf; they are neverthelefs cried up as the greateft fpecifics, and univerfal panaceæ in all dyfenteries, and are even yet frequently prefcribed in them.

It is certain, that fpecifics in general cannot be advifed on any foundation: for moftly the perfon, who recommends them, appeals to experiment alone, in confequence of which an otherwife

wife neglected or dreaded remedy fometimes
appears to do wonderful things in a diforder,
or according to a report often not to be relied
upon, has actually done them.   Experiment is
certainly the beft guide; when he that appeals
to it is capable of making experiments, and
drawing proper conclufions from them.

F I N I S.

E R R A T A.

Page 14. line 13. *for* already, *read* fo early as.
———— l. 15. *dele* out.
P. 28. l. 8. *for* enfues, *r.* it enfues.
P. 30. l. 3. *for* are, *r.* is.
P. 49. l. 19. *for* atribilanious, *r.* atrabilarious.
P. 108. l. 25. *for* premature, *r.* anticipating.
P. 221. l. 26. *for* the patient, *r.* my patient.